Polycyclic Aromatic Hydrocarbons in Water Systems

Authors

David J. Futoma, M.S.
Research Assistant
Department of Chemistry
University of Conneticut
Storrs, Connecticut

S. Ruven Smith, Ph.D.
Professor of Chemistry
University of Connecticut
Storrs, Connecticut

Trudy E. Smith, Ph.D.
Professor of Chemistry
Connecticut College
New London, Connecticut

John Tanaka, Ph.D.
Professor of Chemistry
University of Connecticut
Storrs, Connecticut

CRC Press, Inc.
Boca Raton, Florida

Library of Congress Cataloging in Publication Data
Main entry under titles:

The Analysis of polycyclic aromatic hydrocarbons in
water systems

Bibliography: p.
Includes index.
1. Hydrocarbons--Analysis. 2. Polycyclic compounds--
Analysis. 3. Water--Analysis. I. Futoma, David J.
QD341.H9A56 547'.61046 81-694
ISBN-0-8493-6255-5 AACR2

This book represents information obtained from authentic and highly regarded sources. Reprinted material is quoted with permission, and sources are indicated. A wide variety of references are listed. Every reasonable effort has been made to give reliable data and information, but the author and the publisher cannot assume responsibility for the validity of all materials or for the consequences of their use.

Direct all inquiries to CRC Press, Inc., 2000 N.W. 24th Street, Boca Raton, Florida 33431.

International Standard Book Number 0-8493-6255-5

Library of Congress Card Number 81-694
Printed in the United States

PREFACE

The observation in the early part of this century that certain constituents of coal and its byproducts were tumor producing and potentially carcinogenic failed to touch off any massive degree of scientific inquiry. Analytical investigations on these carcinogenic components, in particular the polycyclic aromatic hydrocarbons, PAH, and their possible proliferation into the environment did not begin in earnest until the early 1960s with the work of Borneff and his associates in Germany. Their analyses of PAH in various water samples were performed primarily by thin layer chromatography after initial solvent extraction and evaporative concentration steps. The number of different analytical methods which have been applied to environmental samples containing PAH as well as synthetic mixtures of PAH from that day to the present day is overwhelming.

The difficulty of effecting quantitative (~99.9%) recovery of PAH from water appears to be reflected in the large number of procedures which have been designed to carry this out. Methods based on various batch and continuous solvent extraction steps are as equally popular as a whole host of extraction procedures utilizing various organic and polymeric adsorbent columns. Most of these techniques are at best empirical and exact relations equating recovery with various parameters such as sample pH, water temperature, ionic strength, etc. are unknown. Clearly then, a need for continued fundamental research is paramount.

In addition to the recovery of PAH, one must be concerned in some cases with raising the concentration of the PAH in the extract to a level detectable by the measuring device employed. Most of the published chromatographic methods can detect nanogram quantities of PAH. Various types of procedures designed to reduce the volume of the extract without loss of solute have been reported but again exact relations governing the processes are lacking.

The next to last step in the analysis has attracted the most attention. This involves the characterization and quantitative determination of each component present. Nearly all environmental samples require an initial separation before the actual measurement or detection step. As a result, the entire range of chromatographic procedures have been employed for separations of PAH, e.g. gas-liquid and gas-solid chromatography with packed columns, capillary column gas chromatography, column and thin-layer chromatography and high performance liquid chromatography. A vast number of different stationary and mobile phases have and are being utilized.

The final step is the actual detection or measurement step. Except for flame ionization, all of these methods have been based on spectroscopic principles. Ultraviolet-visible and infrared adsorption methods, fluorescence and phosphorescence intensity determinations, and mass spectrometric detection comprise the basic core of measurement methods. In efforts to enhance both the sensitivity and the selectivity of the measurement step, a great deal of research is being directed to the development of more sophisticated spectroscopic measurement techniques including measurement of luminescence intensity at cryogenic temperatures.

The preceding outline of analytical techniques points to the need for a review of the methods which have been tested and those which have yet to be exploited. It is clearly not our concern to suggest "the" recommended technique for the analysis of PAH in water. Such a suggestion would in fact be impossible since no one technique is applicable to the whole host of environmental samples possible, samples differing in complexity from relatively "clean" drinking water to "dirty" sludge and sewage. The need as we see it is for a comprehensive classification of the procedures which are available with suggestions as to which methods are best suited for the specific classes of sample which one encounters. Whenever possible, we have singled out optimum methods for a particular separation or measurement. Simple guidelines for future developments have also been offered in all of the areas of research covered in this monograph.

Hopefully, we have succeeded in contributing to the scientific community a handy and, for the most part, thorough review of the modern methods which are and can be used for the analysis of PAH in water systems.

THE AUTHORS

David J. Futoma, M.S., received his B.A. from the University of Connecticut in 1977 and his M.S. degree in Alalytical Chemistry from the same institution in 1980. His professional experience includes undergraduate teaching assistant from 1976 to 1977 and graduate research assistant from 1978 to 1980.

Mr. Futoma is a member of the American Chemical Society, Phi Lambda Upsilon, Phi Kappa Phi, and Phi Beta Kappa.

His present research interests center around the analyses of environmental samples for carcinogenic and other potentially harmful pollutants. Along this same theme, much of his present efforts are focused on improving public awareness of existing environmental hazards as well as introducing environmental topics into secondary school science programs.

S. Ruven Smith, Ph.D., earned his B.Sc. degree in Chemistry and Physics from McMaster University in 1942 and his M.Sc. degree in Physical Chemistry from the same institution in 1943. He spent from 1943 to 1946 as a junior research chemist with the National Research Council of Canada. He received his Ph.D. degree in Physical Chemistry from Ohio State University in 1952. From 1952 to 1960 he was a research chemist at the Naval Weapons Center at China Lake, California. From 1960 to the present he has been at the University of Connecticut and a consultant at the Naval Weapons Center, China Lake, CA. His research interests include mass spectroscopy, mass spectrometric analyses, reaction kinetics, and pollution studies.

Trudy E. Smith, Ph.D., earned her B.A. degree in Chemistry and Biology from Greenboro College in 1944 and her Ph.D. degree in Physical Chemistry in 1957 from Ohio State University. From 1951 to 1953 she was a research chemist at Aerojet-General Corporation in Azusa, CA and from 1953 to 1960 she was a research chemist at the Naval Weapons Center, China Lake, CA. From 1960 to 1962 she was at the University of Connecticut and since 1962 has been at Connecticut College, New London, CT. She is presently Chairman of the Department of Chemistry. Her research interests include gas phase reaction kinetics and mass spectrometry.

John Tanaka, Ph.D., received his undergraduate degree in chemistry at the University of California at Los Angeles in 1951 and his Ph.D. degree in chemistry from Iowa State University in 1956. After spending 1956 to 1963 teaching at South Dakota State College, he took a 2-year post-doctoral at the University of Pittsburgh. He has been at the University of Connecticut since 1965. Tanaka's research interests have spanned a broad spectrum and have ranged from inorganic hydrides to electrical insulating materials. One of the continuing strands in this spectrum of research projects has been an interest in analysis of organic and metal-organic compounds.

ACKNOWLEDGMENT

The work upon which this publication is based was supported in part by funds provided by the Office of Water Research and Technology (Project No. A-084-CT) U.S. Department of the Interior, Washington, D.C. as authorized by the Water Research and Development Act of 1978 (P.L. 95-467).

We wish to thank Professor John T. Stock, Department of Chemistry, University of Connecticut for reading the manuscript.

The sincere devotion of Ms. Linda Colburn, Ms. Lane Witherill, and Ms. Bonnie Crowe in preparing the manuscript is gratefully acknowledged. Special acknowledgment is extended to Ms. Lori Krol who gave of herself assistance above and beyond the normal call of duty.

TABLE OF CONTENTS

Chapter 1

INTRODUCTION

Polycyclic aromatic hydrocarbons, PAH, are homologs of benzene in which three or more aromatic rings are joined in different configurations. Some common PAH and their abbreviated names are listed in Table 1.

It has been known for many years that certain PAH possess carcinogenic properties. In a paper presented at a Meeting of the Analytical Division of the London Chemical Society in February of 1978, Roe[1] cites a reference dating back to 1932 linking the incidence of cancer to benzo(*a*)pyrene, B(*a*)P, and dibenz(*a,h*)anthracene. This is well before analytical studies of PAH in water began. Roe's paper presents a brief outline of the cancer activity studies of PAH. Table 1 summarizes the carcinogenic potentials of several PAH. Recently, in a volume edited by Gelboin and Ts'o,[2] a number of aspects of the carcinogenesis studies on PAH have been presented. In addition, Volumes 1 and 3 of *Carcinogenesis — A Comprehensive Survey*[3,4] cover PAH and their chemistry, metabolism, and carcinogenic effects. These surveys deal with a few analytical studies of PAH in the environment but tend to stress the biological effects of PAH on man. Most studies on the carcinogenesis of PAH have been carried out with B(a)P. Although B(a)P and other PAH have been shown to be strongly carcinogenic, it has never been definitively determined that the levels of PAH found in drinking water are harmful. The World Health Organization in Europe recommends that the total concentration of six specific PAH not exceed 200 ng/ℓ for domestic drinking water.[5] The six specified PAH are fluoranthene, benzo(*b*)fluoranthene, benzo(*k*)fluoranthene, benzo(*a*)pyrene, benzo(*ghi*)perylene, and indeno(1,2,3-cd)pyrene. Specific limits have not been set in the U.S. but the EPA has recognized several PAH as priority pollutants in water.[5a]

A number of comprehensive reviews have covered the literature dealing with the occurrence and significance of PAH in the aqueous environment.[6-9] Andelman and Suess[6] review the literature of the 1960s and the pioneering efforts of Borneff et al. on PAH analyses. In a later review, Andelman and Snodgrass[7] cover the literature from 1968 to 1973 discussing the sources and origins of PAH. They present numerous tabulated values of PAH concentrations for various aqueous environmental samples. About the same time, Hutzinger et al.[8] published a literature review which summarily covers the occurrence and analysis of PAH in various environmental samples, however, problems of PAH in the aqueous environment are not stressed. In a comprehensive review published by Harrison et al.[9] in 1975, essentially all aspects of PAH in water are discussed. The review contains numerous references to the sources and occurrences of PAH. It includes many tables of values of PAH concentrations in numerous water samples. The tables also show the efficiencies of removal of PAH from water by various decontamination methods. Their review covers both the solubility and analysis of PAH in water, but all the solubility data dates from the 1950s and analytical techniques covered are not representative of modern methods.

Biennially, reviews of water pollution studies appear in *Analytical Chemistry,* with occasional mention of PAH. The most recent by Fishman and Erdmann[10] presents only some selected literature of PAH in water and is not very useful to the analytical chemist who is interested in a detailed knowledge of PAH analyses. Reviews of analyses of organics in water appear in *Journal of the Water Pollution Control Federation* annually with a fairly thorough listing of literature articles relating to PAH in water. However, there are few details on analytical techniques and the review is not intended to be critical. The most recent review on organic pollutants in water in the *Journal of the Water Pollution Control Federation* is by Chian and DeWalle.[11] In general, most reviews on water pollution do not place particular emphasis on PAH.

Table 1
PAH AND OTHER AROMATIC HYDROCARBONS

Structure	1957 I.U.P.A.C. name	Other names	Mol wt	Relative carcinogenicity[a,b]	Common liter. abbrev. (if any)
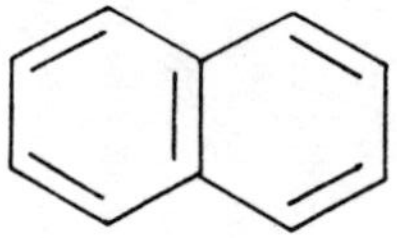	Naphthalene	—	128	—	—
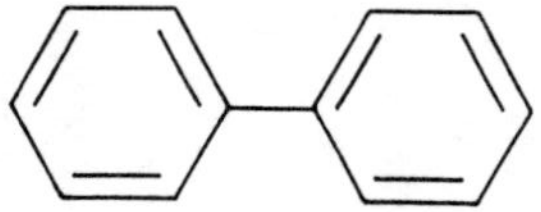	Biphenyl	—	154	—	—
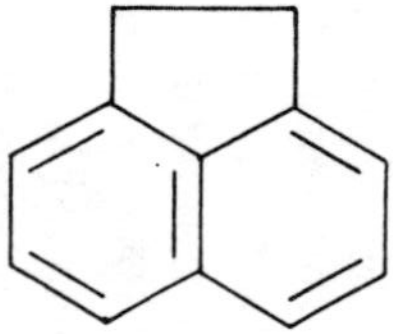	Acenaphthene	—	154	—	—
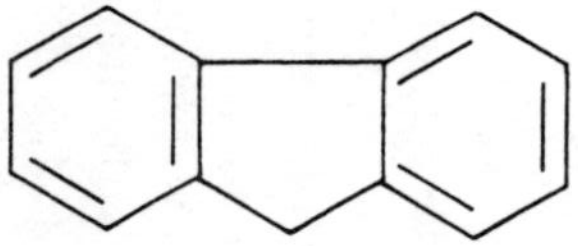	Fluorene	—	166	—	—
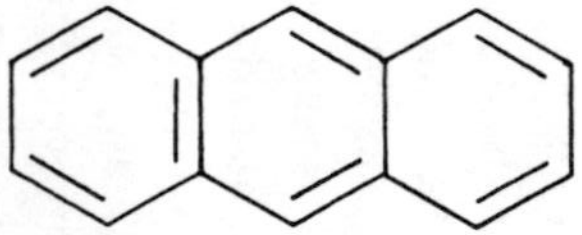	Anthracene	—	178	—	—

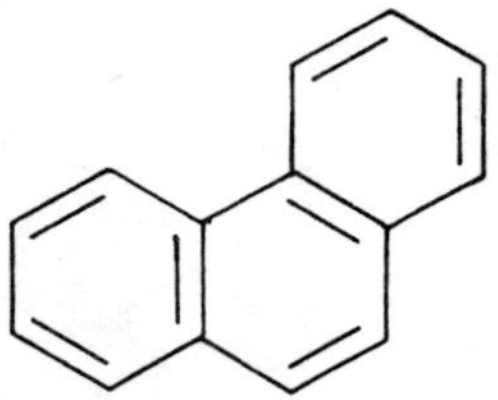	Phenanthrene	—	178	—	—
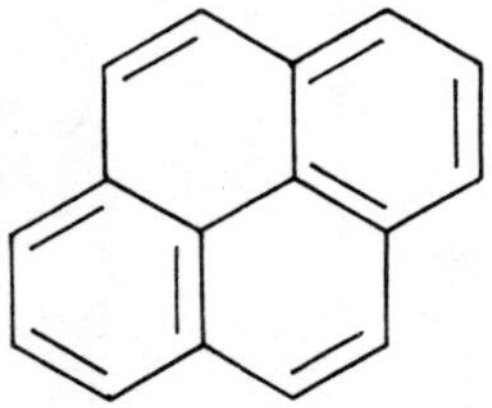	Pyrene	—	202	—	—
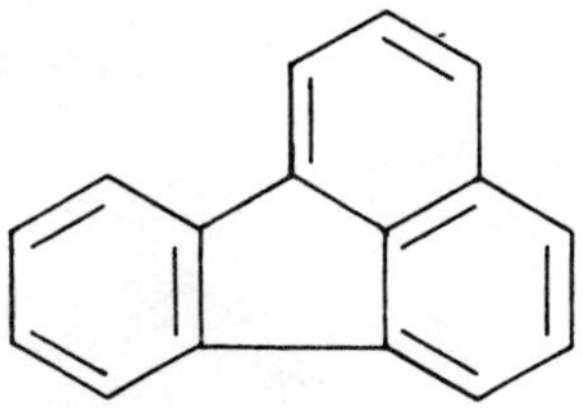	Fluoranthene	—	202	—	—
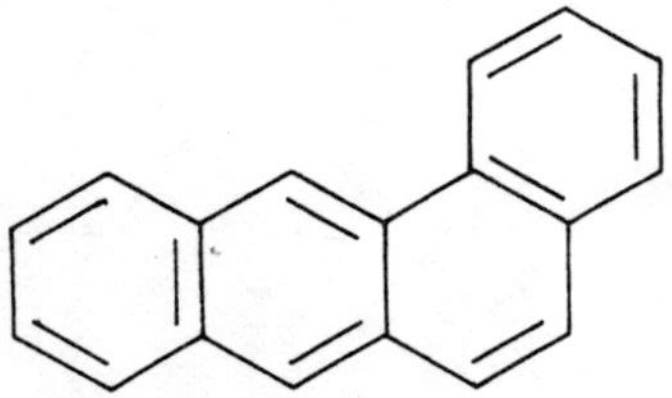	Benzo(a)anthracene	1,2 Benzanthracene	228	<+	B(a)A

Table 1 (Continued)
PAH AND OTHER AROMATIC HYDROCARBONS

Structure	1957 I.U.P.A.C. name	Other names	Mol wt	Relative carcinogenicity[a,b]	Common liter. abbrev. (if any)
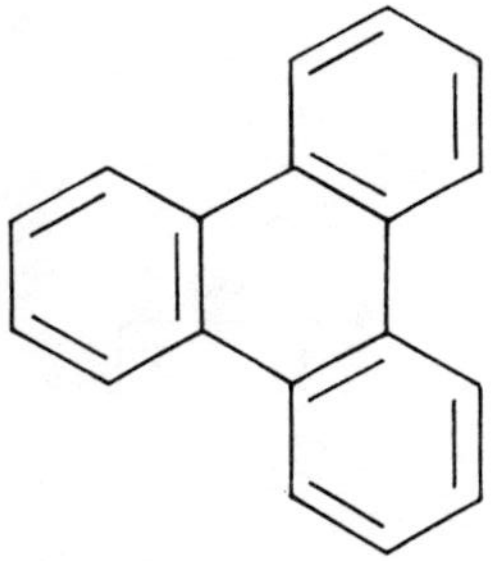	Triphenylene	—	228	—	—
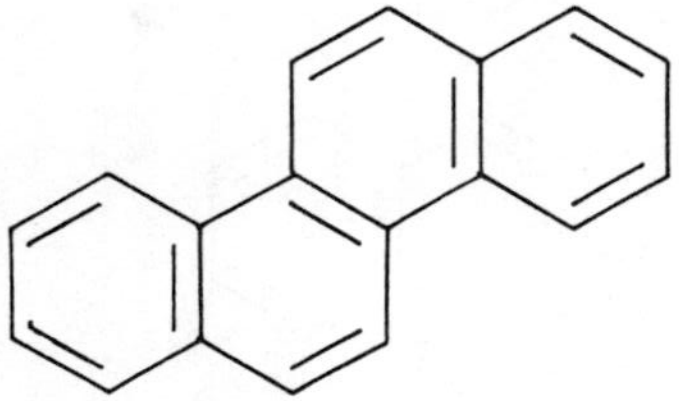	Chrysene	—	228	<+	—
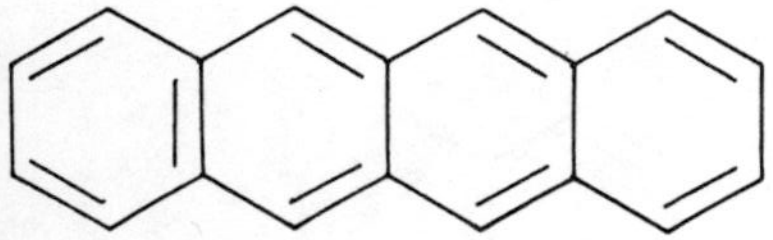	Naphthacene	Tetracene	228	—	—

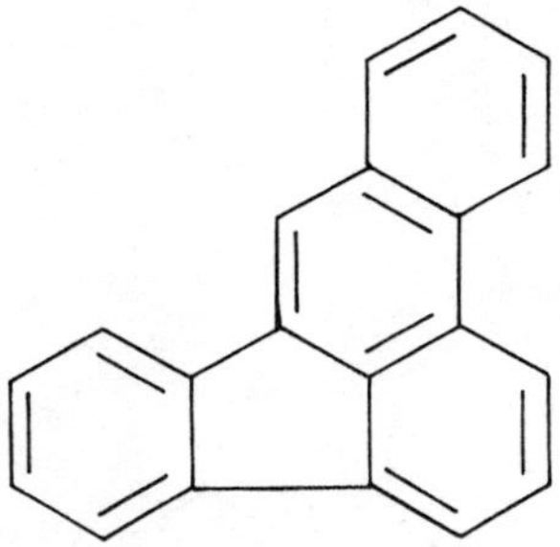	Benzo(b)fluoranthene	11922 3,4 Benzfluoranthene	252	+ +	B(b)F
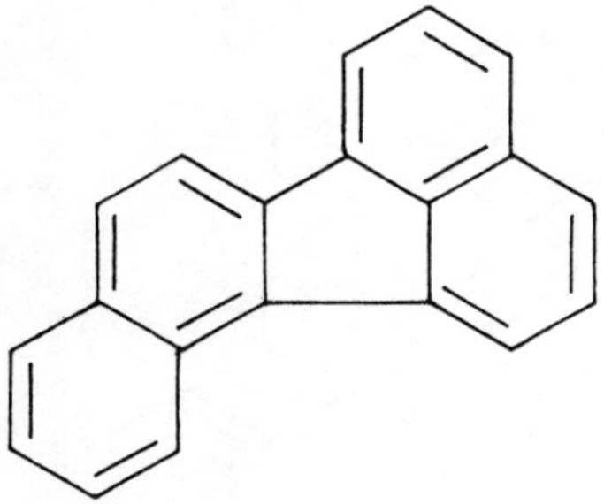	Benzo(j)fluoranthene	15938 10,11 Benzfluoranthene	252	+ +	B(j)F
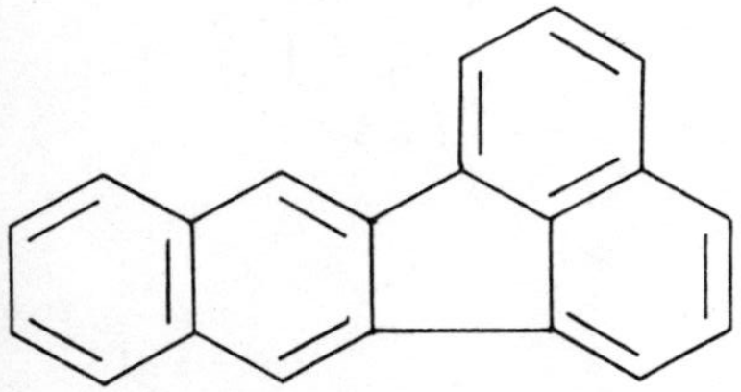	Benzo(k)fluoranthene	11,12 Benzfluoranthene	252	—	B(k)F

Table 1 (Continued)
PAH AND OTHER AROMATIC HYDROCARBONS

Structure	1957 I.U.P.A.C. name	Other names	Mol wt	Relative carcinogenicity[a,b]	Common liter. abbrev. (if any)
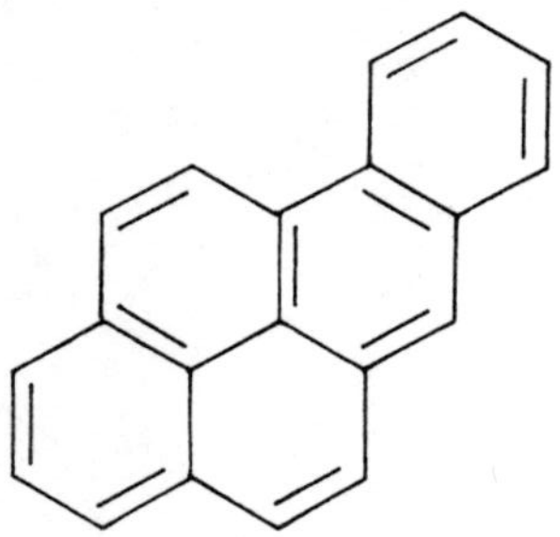	Benzo(a)pyrene	3,4 Benzopyrene	252	+ + + +	B(a)P
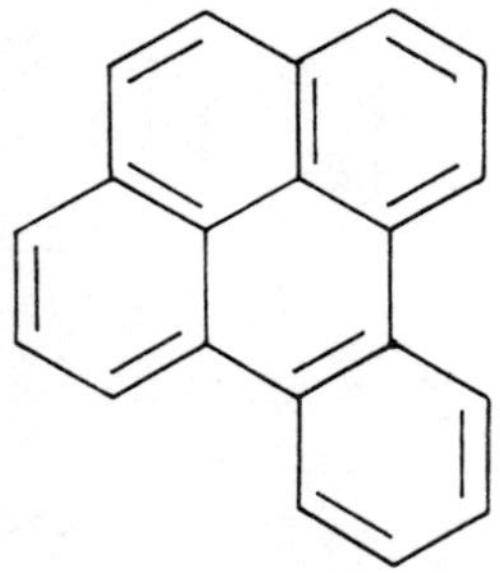	Benzo(e)pyrene	1,2 Benzopyrene	252	<+	B(e)P

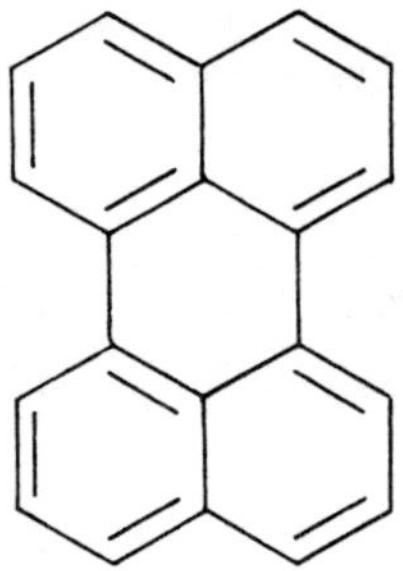	Perylene	—	252	—	—
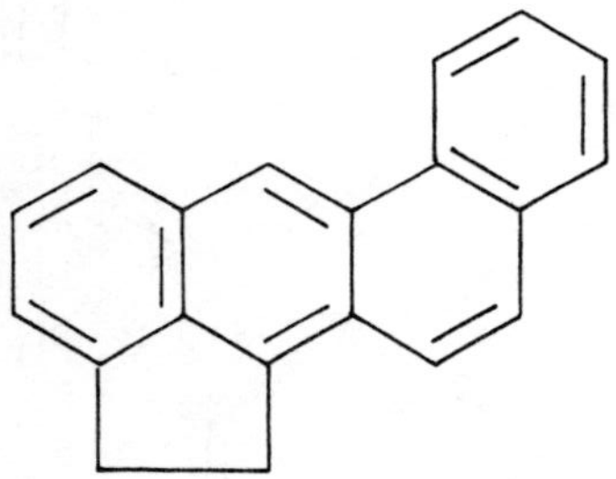	Cholanthrene	—	254	—	—
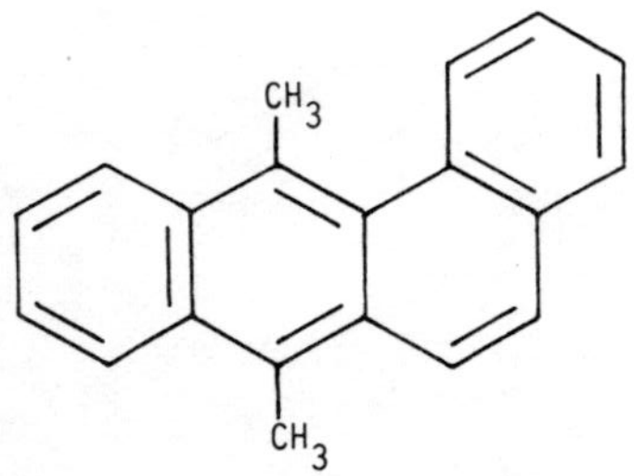	7,12 Dimethylbenz-(a)anthracene	7,12 Dimethyl-1,2-benzanthracene	256	+ + + + +	—

Table 1 (continued)
PAH AND OTHER AROMATIC HYDROCARBONS

Structure	1957 I.U.P.A.C. name	Other names	Mol wt	Relative carcinogenicity[a,b]	Common liter. abbrev. (if any)
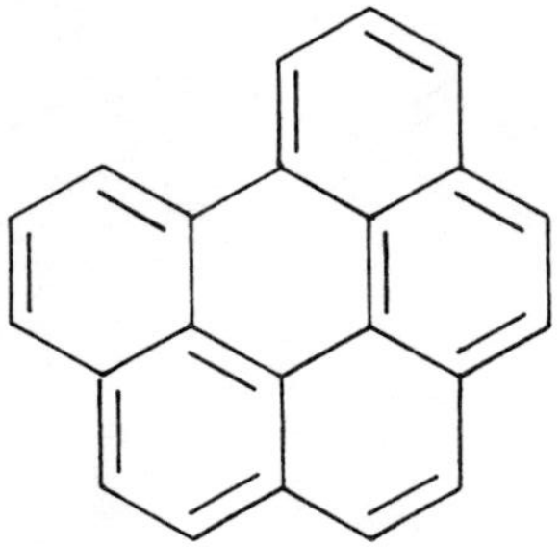	Benzo(ghi)perylene	1,12 Benzperylene	276	—	—
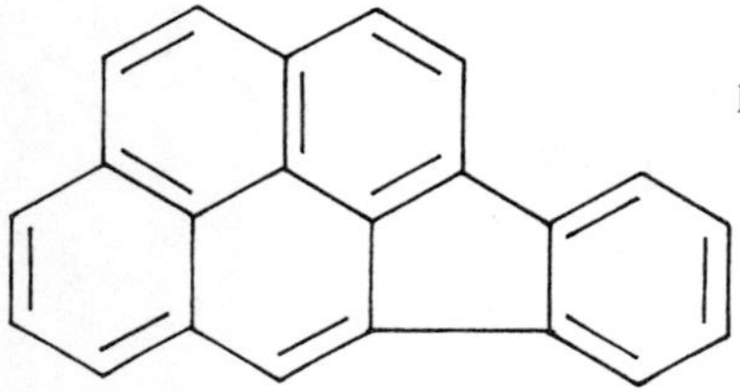	Indeno(1,2,3-cd)pyrene	o-Phenylenepyrene	276	+	IP
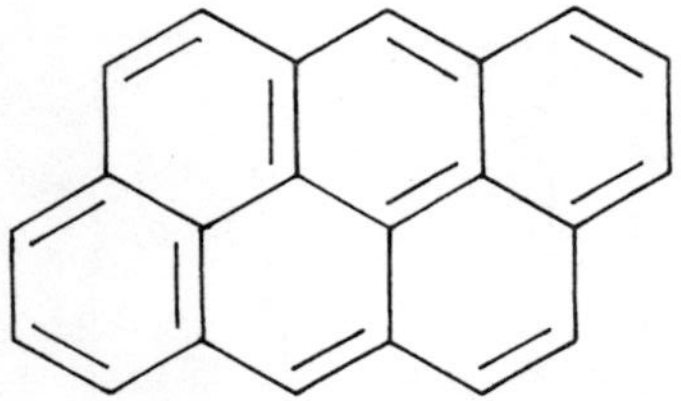	Anthanthrene	—	276	<+	—

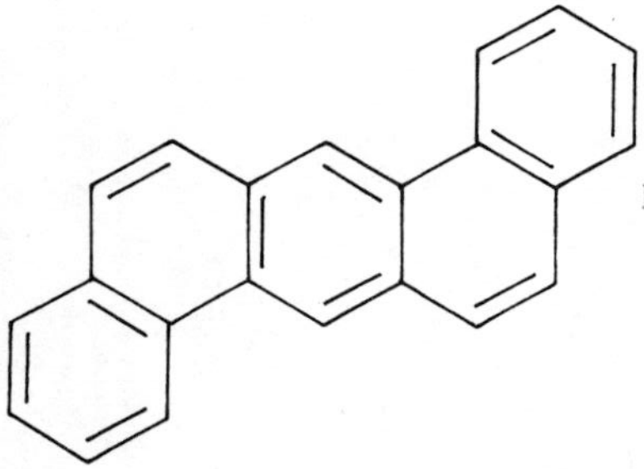	Dibenz(a,h)anthracene	1,2,5,6 Dibenzanthracene	278	+ + +	—
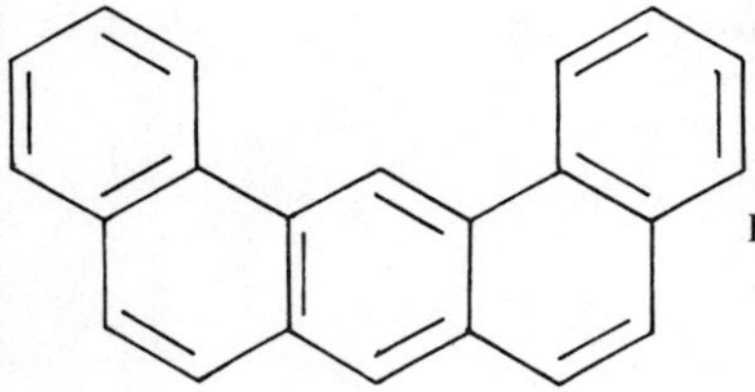	Dibenz(a,j)anthracene	1,2,7,8 Dibenzanthracene	278	—	—
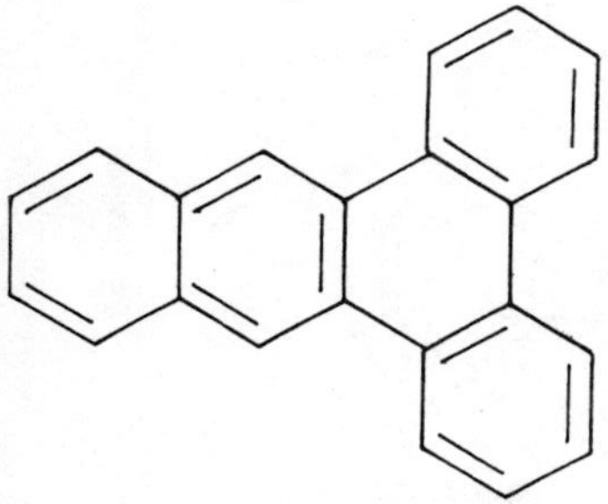	Dibenz(a,c)anthracene	1,2,3,4, Dibenzanthracene	278	—	—

Table 1 (continued)
PAH AND OTHER AROMATIC HYDROCARBONS

Structure	1957 I.U.P.A.C. name	Other names	Mol wt	Relative carcinogenicity[a,b]	Common liter. abbrev. (if any)
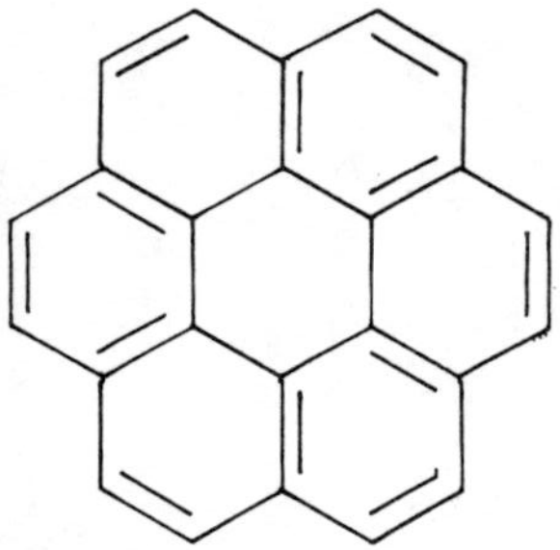	Coronene	—	300	—	—

[a] Carcinogenicity values taken from reference 4, pp. 278-281 and reference 9.
[b] + + + + + = extremely active, + + + + = very active, + + + = active, + + = moderately active, + = weakly active, < means less than, − = inactive or unknown.

In the past 5 years, analytical methodology applied to PAH systems has undergone rapid growth and development. Multicomponent mixtures of PAH can be more effectively separated and detected at lower levels of concentration. Improvements in selectivity of the stationary phases used in gas chromatography (gc) and high pressure liquid chromatography (hplc) have contributed to the improvements in qualitative analytical studies. Glass capillary (gc) has also been shown to be a powerful technique for separating PAH in complex mixtures. The development of highly sensitive fluorescence detectors for hplc has reduced the levels of PAH detection down to the picogram range. The combination of gc and glass capillary (gc) to mass spectrometry has also been developed, providing sub-nanogram detection of PAH.

These chromatographic advances have been complemented by several innovative spectroscopic analytical techniques. Meant to be both qualitative and quantitative, they include matrix isolation Fourier Transform infrared spectroscopy, (IR), synchronous luminescence spectroscopy, wavelength modulation and derivative spectroscopy, low temperature fluorescence spectroscopy, and room temperature phosphorescence spectroscopy. Research on the above mentioned techniques is in the developmental stage but significant advances have been made in the last few years.

From 1976 to the present, a number of solubility studies of PAH in water by novel techniques have appeared. Methods of preconcentration have also been more fully investigated in the last few years. Although there have been several comprehensive reviews concerning the incidence and significance of PAH in the environment, no comprehensive review has been published on the various state of the art analytical techniques dealing with environmental samples of PAH. There clearly exists a need to evaluate all the work that has been done and decide on the most suitable approaches that can be employed for analyses of the complex samples of PAH which occur in the environment.

Chapter 2

SOLUBILITY STUDIES OF PAH IN WATER

I. GENERAL SOLUTION THEORY

Solution theory applied to solubility studies of PAH in water can be useful in its ability to make general predictions and estimates of solubility values. Hildebrand and Scott,[12] in their classic treatment on the solubility of nonelectrolytes, summarized the theoretical background upon which numerous attempts to correlate hydrocarbon solubilities with thermodynamic parameters were made. However, many solubility values are still being determined experimentally because of the continuing validity of their comment that, "Indeed, the sober-minded worker in this field is likely to come to regard as utopian any hope of predicting solubilities from the properties of the pure components to better than, say 10%. If one must know a solubility to 1%, he should measure it, and that with some care."[12]

Solubility theory of PAH finds its roots in the solubility studies of benzene, naphthalene, and their alkyl derivatives. In part, the higher water solubility of these substances made experimental determinations easier. Also the interest 30 years ago in PAH was somewhat less than today. One of the earlier proposed theories,[13] which has remained essentially unrevised in modern studies, considers two major mechanisms responsible for the solubility of aromatics (e.g., benzene and naphthalene) in water. The positive heat of solution term which acts against increasing solubility involves a process of breaking up the water structure to create a cavity for the hydrocarbon to enter into solution. The negative heat of solution term, which must have some contribution in order for hydrocarbons to dissolve at all, arises from two sources. One conjecture is that the solution process can in some manner rearrange the water structure to create more order through a hydrogen-bonded network around the hydrocarbon molecule. Another contributor to the solubility enhancing term might involve an interaction, most likely weak, between aromatic pi electrons and the water hydrogens.

If this proposed theory is valid, the decreasing solubility of PAH in water as the number of rings increases could be explained by considering the "cavity formation" process to be energetically more unfavorable than the gain in attractive energy by adding more pi electrons. The observation of a small solubility decrease due to added alkyl groups could simply be a size effect.

These general theories are translated into specific measurements which can be used to predict the solubility of PAH in water. Thermodynamic parameters are extensively used. Experimentally available terms such as molar volumes and activity coefficients are related to solubilities. Variables such as carbon number, i.e., the number of carbon atoms in a molecule, are correlated with solubilities.

Early correlation efforts[14,15] focused mainly on easily obtainable parameters, e.g., molar volume with a qualitative discussion of carbon chain structure and alkyl substituent effects. A thermodynamic framework for correlating solubilities began with the determination by Tsonopolous and Prausnitz[16] of over 100 activity coefficients of aromatic solutes.[16] The general approach has been to correlate the hydrocarbon infinite dilution activity coefficient, γ_ω^∞ with carbon number. For liquid hydrocarbons, γ_ω^∞ is approximately equal to the finite dilution activity coefficient with the final result that γ_ω^∞ is just the reciprocal of the mole fraction solubility.[17] The treatment for solid PAH, i.e., those with melting points above 25°C, is somewhat more complex. The theoretical solubility involves some estimates of experimentally inaccessible quantities.[17] In the equation,

$$\gamma_\omega = F/x_\omega \qquad (1)$$

γ_ω is the finite dilution activity coefficient, χ_ω is the mole fraction of the PAH in water, and F is the ratio of the fugacity of the solid hydrocarbon to the fugacity of the pure subcooled liquid hydrocarbon at the system temperature. The latter fugacity is experimentally inaccessible and the following approximation is made

$$\ell nF = K(T_{tp} - 298.15) \qquad (2)$$

where T_{tp} is the triple point temperature of the hydrocarbon and K is a constant. The equation is based on an assumption that all PAH have approximately the same entropies of fusion. From available data, K averages out to be about −0.023. This value is used to arrive at F for PAH not experimentally studied. Using experimentally determined solubilities along with F, values of γ_ω can be tabulated. Values of γ_ω for experimentally studied PAH are then plotted vs. carbon number which allows γ_ω to be found for any PAH. This plot shows some large deviations and these deviations are particularly serious for alkyl substituents and very high molecular weight PAH. If such parameters as the molar volume of the subcooled hydrocarbon were available, a more linear plot might result. Since the only easily obtainable parameter is carbon number, best fit methods are applied and the resulting equation is

$$\log\gamma_\omega^\infty = 3.5055 + 0.3417\,(N\text{-}6) - 0.002640\,(N\text{-}6)^2 \qquad (3)$$

where N is the number of carbon atoms.[17] Using Equation 3 along with 2, the solubility can be calculated from Equation 1. The correlation extends over five orders of magnitude and involves prior knowledge of only triple point temperature and carbon number. Accuracies never approach the "utopian 10%" but for crude estimates good to a factor of two or three the equation is useful. Figure 1 compares correlated and experimental solubilities.

Some improvements in methods of experimentally obtaining thermodynamic parameters might reduce the error associated with these estimates, yet an experimental solubility calculation is probably the only way to obtain an accuracy with an error of less than 1 to 5%. However, these largely empirical correlations become attractive when it is possible to estimate a solubility which is otherwise obtainable only with a great deal of experimental difficulty.

II. EXPERIMENTALLY DETERMINED SOLUBILITIES

A. Preparation of a Saturated Solution of PAH in Water

When an estimate of solubility from theoretical considerations will not suffice, an experimental determination is necessary. Since the aqueous solubilities are very low for PAH, (see Table 2), solubility studies are concerned with only saturated solutions. Working with extremely dilute solutions presents some serious problems such as adsorption on container and cell walls, accomodation or the existence of colloidal or micellar particles, interference effects from dust or other impurities, and the reproducibility of preparing such solutions.

The classical and still common method for preparing saturated PAH solutions involves adding an excess amount of the solid to water, agitating by stirring or mechanical shaking, and allowing dissolution to occur over a period of time. Numerous solu-

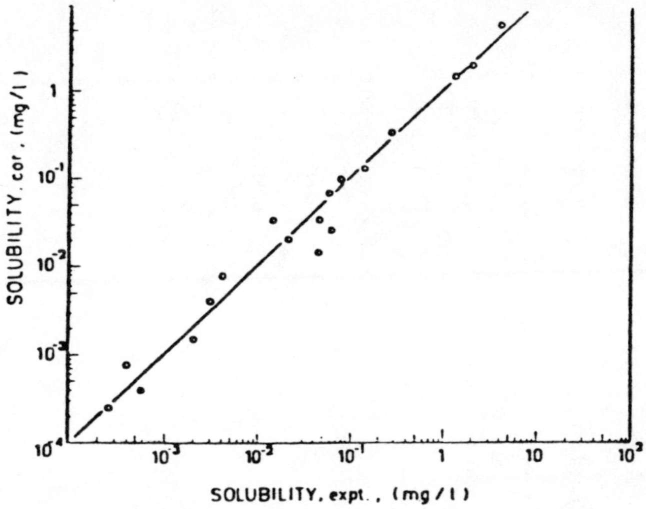

FIGURE 1. Comparison of correlated and experimental solubilities for PAH. (Reprinted with permission from Mackay, D. and Shiu, W. Y., *J. Chem. Eng. Data*, 22, 399, 1977. Copyright by the American Chemical Society.)

bility studies have employed this technique.[13,17-22] In all cases, some sort of temperature control is used, generally at 25°C for room temperature studies. The length of time allowed for equilibration is not standardized and 24 hr,[22] 48 hr,[12,13,17] 4 to 7 days,[19,20] and 1 to 3 weeks[18] have all been reported. An equilibration time is definitely needed since excess PAH must be allowed to settle out. Once the solution is considered to be saturated there are a number of available methods which have been devised to allow samples to be taken for analysis. A siphoning glass tube delivery system[13] and pipette withdrawal through a glass wool plug[18] have been used. After decantation and filtration through a 5μm Millipore® filter,[17] the PAH solution is ready for extraction. Another more elaborate approach involves a special design feature in the flask employed to dissolve the PAH.[20] Halfway up a Pyrex® cylinder capped by a Pyrex® ground glass stopper, a protruding Pyrex® tube some several centimeters long is attached. The tube can be sealed by a stainless steel Swagelok® cap and samples are taken from a region of solution not affected by interfering floating or sedimented hydrocarbon particles. Once a sample is taken, a suitable solvent for the preconcentration prior to solubility determination may be needed. Cyclohexane is a common solvent used for the extraction step prior to spectrophotometric analysis. A large volume of aqueous solution can be reduced down to a few milliliters of organic solvent, allowing the absorption or fluorescence process to occur in a more favorable concentration range, e.g., mg/ℓ.

Another method of preparation for saturated PAH solutions by May et al.[23-25] is based on a dynamic process whereby water is forced through an exchange-type column, called a generator column, containing glass beads coated with the particular compound to be studied. The stainless steel column is 60 × 0.6 cm (see Figure 2) packed with 60 to 80 mesh glass beads coated with 1% w/w of the particular PAH. The coating is applied by addition of 20 g of glass beads to 200 mℓ of a 0.1% solution by weight of the PAH in methylene chloride, and then evaporating off the solvent. Using a temperature controlled system, solutions are prepared by pumping water at a speed of no

Table 2
COMPARISON OF SOLUBILITY DATA OF PAH BY SEVERAL INVESTIGATORS

Formula & compound	May et al.[24] (1978)	Schwarz[20] (1977)	Mackay & Shiu[17] (1977)	Wauchope & Getzen[18] (1972)	Klevens[15] (1950)	Davis[26] 27°C (1942)
Naphthalene $C_{10}H_8$	31.69 ± 0.23	30.3 ± 0.3	31.7 ± 0.2	31.2		
Fluorene $C_{13}H_{10}$	1.685 ± 0.005		1.98 ± 0.04	1.90		
Anthracene $C_{14}H_{10}$	0.0446 ± 0.0002	0.041 ± 0.0003	0.073 ± 0.005	0.075	0.075	0.075 ± 0.005
Phenanthrene $C_{14}H_{10}$	1.002 ± 0.011	1.151 ± 0.015	1.290 ± 0.07	1.180	1.600	1.600 ± 0.05
Fluoranthene $C_{16}H_{10}$	0.206 ± 0.002		0.260 ± 0.02	0.265	0.265	0.240 ± 0.02
Pyrene $C_{16}H_{14}$	0.132 ± 0.001	0.129 ± 0.0002	0.135 ± 0.005	0.148	0.175	0.165 ± 0.007
Benz (*a*) anthracene $C_{18}H_{12}$	0.0094 ± 0.0001		0.014 ± 0.0002		0.010	0.011 ± 0.001
Chrysene $C_{18}H_{12}$	0.0018 ± 0.0001		0.002 ± 0.0002		0.006	0.0015 ± 0.0004
Perylene $C_{20}H_{12}$			0.0004 ± 0.00002			0.0005 ±
Benzo (*a*) pyrene $C_{20}H_{12}$			0.0038 ± 0.00031			0.004 ±
Coronene $C_{24}H_{12}$			0.000414 ± 0.00002			

Note: All solubilities in (mg/kg) and at 25°C except where noted.

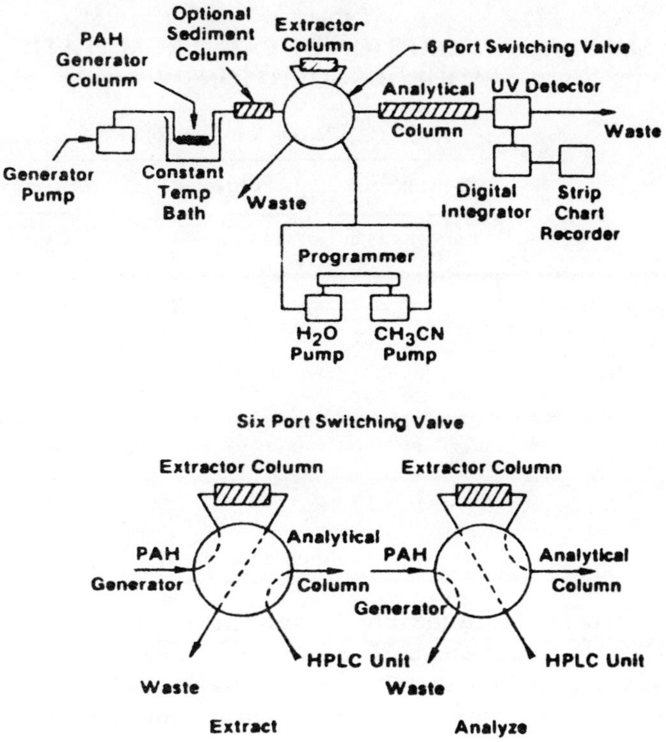

FIGURE 2. Dynamic coupled-column liquid chromatographic flow diagram.

greater than 5 mℓ/min through the column. The extraction system employed does not involve the usual separatory funnel but rather the passage of a predetermined amount of the saturated solution through an extractor column packed with a superficially porous bonded C_{18} stationary phase, (Bondapak® C_{18}, Waters Associates). An extraction efficiency of 99% is claimed.[23] After the column extraction step, an acetonitrile-water solvent blend is used to elute the PAH from the octadecyl, C_{18}, stationary phase. The next step, unique to this technique, is actually a cleanup procedure whereby the solution of PAH is passed through a 30 × 0.6 cm microparticulate analytical column (μ-Bondapak® C_{18}, Waters Associates) which separates the PAH from interfering substances. To minimize band broadening when necessary, a linear solvent gradient is employed. The PAH solution which now contains only the substance being measured is ready for detection by an UV method.

Both methods of solution preparation have several sources of error in common though their relative magnitudes may differ. The principal disadvantages of the static saturation and classical sampling method include adsorption of PAH on the solution vessel walls, pipettes and filter paper, incomplete equilibration of the PAH, as well as the formation of colloidal or particulate PAH in the water, i.e., accomodation. Table 3 lists some of the adsorption characteristics of three PAH demonstrating the significance of the problem. The errors due to adsorption effects are minimized in the dynamic coupled column liquid chromatographic DCCLC technique. In the dynamic system the contact area where PAH can adsorb is small with only 6 $\mu\ell$ of volume between the generator and extractor columns. With a 5mℓ/min flow rate, the time of passage between solution preparation and extraction will be on the order of milliseconds. To

Table 3
SURFACE ADSORPTION OF PAHs FROM STATIC AQUEOUS SOLUTIONS[1]

	Percent loss from solution					
	Phenanthrene		Chrysene		Benz[*a*]pyrene	
	1 hr	13 hr	1 hr	13 hr	1 hr	13 hr
Glass	8		46	71	53	82
Silanized glass	5	73	64	81	73	93
Platinum	35	87	66	85	57	93
Aluminum	5	76	50	89	67	95

[1] The concentration of each PAH was 1 μg/kg. Surface to volume ratios are approximately 1 cm^2/cm^3. All measurements are made by the coupled column LC technique.

further prevent adsorption, the transfer lines are presaturated with the PAH under study.

Adsorption effects precluded the direct introduction of the saturated aqueous solutions into the UV detection system by a sample loop. Table 4 indicates that the concentration measured depends on the volume of saturated solution which has passed through the loop before the injection, indicating adsorption on the loop walls.

Another potential error which must be considered in the saturation process concerns the formation of colloidal and particulate PAH. The extent of accomodation (difference between the measured and equilibrium amounts of hydrocarbon in water) depends on several factors: the amount of hydrocarbon present, i.e., the hydrocarbon supply, the time allotted for the particles to settle out, the pore size of the filters employed, and the manner in which the hydrocarbons are introduced into the solutions.[23] It is not clear whether most investigators were aware of this possible problem but May et al.[23-25] are the only ones who have considered the effect.

In order to verify that the DCCLC approach was not liable to produce systematic errors due to accomodation, May et al.[23] devised a few experimental tests. To discount the effect of the amount of hydrocarbon used to prepare the solution, an alternative approach was adopted for preparing saturated PAH solutions whereby water was pumped through a column loaded with the crystalline solid phenanthrene rather than phenanthrene supported on glass beads. Solution concentrations were not affected by this alternative procedure as shown in Table 5. Solutions of invariant concentration were obtained even when large volumes of water were pumped through the generator columns, since the amount of PAH going into solution was only a minute fraction of the material on the generator column. This effect is demonstrated in Table 6. Since the system was dynamic and the time lag between solution preparation and detection was very small, settling times were considered to have no effect.

Filtration did not cause a concentration change in the solutions produced by the generator column. Since filters are a part of the column, fitted onto both ends, the filters are saturated with the PAH under study when the column is conditioned and can be considered to be a major part of the generator column.

In the classical approaches, the adsorption effect is the only problem addressed. For the most part, it is considered to have a minimal effect. The solution sampling process through the side of the cylinder mentioned earlier[20] was claimed, without detailed discussion, to minimize losses due to adsorption. In one experiment, spectroscopic eval-

Table 4
ADSORPTION OF PHENANTHRENE FROM AQUEOUS SOLUTION TO STAINLESS STEEL SAMPLE LOOP

Volume (mℓ) of 0.95 ppm phenanthrene solution[a] passed through sample loop[b]	Apparent concentration, ppm
0.1	0.97
4.0	1.36
9.2	1.61
11.0	1.70

[a] Known phenanthrene concentration, 0.95 ± 0.02 ppm.

[b] Sample loop volume, 23.1 μℓ. Phenanthrene flow rate, 0.4 mℓ/min.

Table 5
SOLUBILITY DEPENDENCE ON THE SUPPLY OF PHENANTHRENE ON A GENERATOR COLUMN

1% Phenanthrene on glass beads column (60 × 0.6 cm) (μg/kg)[a]	Crystalline phenanthrene column (60 0.6 cm) (μg/kg)[a]
959	938
947	935
958	943
Av 935 ± 7	Av 939 ± 4

Note: Aqueous flow rate, 5 m/min. Temperature, 24.3°C.

[a] Phenanthrene concentration in generator column effluents.

uation of a second extraction step of the aqueous saturated solution indicated complete extraction efficiency[18] and hence adsorption effects were considered minimal. In the one experiment where filtration was followed by extraction,[21] adsorption effects were considered negligible since the quantities of solution were so large that differences in concentration brought about by adsorption would be unnoticeable. Results in this case were identical for a nonfiltration technique.

Which method of preparation of saturated solutions seems to be best? A quick glance at Table 2 shows little difference in solubilities measured between May[23] and Schwarz,[20] where the DCCLC approach and the slow equilibration method with a solution vessel modified with a side arm delivery tube are used, respectively. The advantages of the DCCLC technique stem from the circumvention of a manual sampling and solution process. The dynamic nature of the procedure seems to minimize adsorp-

Table 6
EFFECT ON THE SOLUBILITY MEASUREMENT OF PASSING LARGE VOLUMES OF WATER THROUGH THE GENERATOR COLUMN

Volume of distilled water eluted through column, m*l*	Concentration measured, mg/kg
1	45.2
190	47.0
230	46.6
500	46.1
775	46.9
910	44.6
1175	45.5
Average measured concentration (6-21-77)	46.0 ± 0.9 μg/kg

tion and accommodation difficulties. The use of a "cleanup' column will also tend to remove interfering solutes, thus making the eventual detection step easier and more accurate.

Besides the relative simplicity of preparing saturated solutions by the slow equilibration approach, the one main advantage to this approach is that it gives results comparable to the DCCLC method. A deliberate and careful analytical lab technique should be strictly followed when manually preparing these solutions, since adsorption effects are potentially more serious due to the increased contact time.

B. Analytical Determination of Solubilities

Nephelometry,[26] gas chromatography,[14] UV spectrophotometry,[13,18,22] fluorescence spectrophotometry,[17,19-21] and UV detection in conjunction with the DCCLC technique[23-25] have all been employed in the analytical determination of PAH solubilities.

Nephelometry is the oldest method but it has been supplanted by more modern techniques due to its relative lack of sensitivity compared to UV and fluorescence detection.

Gas chromatography is limited to a few lower molecular weight aromatic hydrocarbons with sufficiently high volatilities, such as naphthalene, biphenyl, and their alkyl derivatives.

In a typical ultraviolet determination,[18] the saturated aqueous solution is extracted with cyclohexane. The volume of the solvent is then adjusted to allow absorbance readings to fall in the range of 0.5 to 1.5. Standard PAH solutions in cyclohexane are prepared and serially diluted such that a series of six or more absorbance readings can be made to bracket the 0.5 to 1.5 absorbance range. Measurements are made on a Cary® 14 spectrometer. Experimental Beer's Law errors of 0.5% are noted, and reproducibilities are on the order of 2 to 5%. Table 2 lists some of the solubility results of the method and in Table 7 the Beer's Law parameters and concentration ranges of the PAH are studied.

For the fluorescence method, the approach used by Schwarz[20,27] is representative of the state of the art. By adding a quartz tube to the top of standard quartz fluorescence cells, each cell can be sealed with a stainless steel Swagelok® cap, thus preventing evaporation. Thermal equilibration of the saturated solutions is not assumed until the sealed cell is rotated for 3 days in a temperature-controlled water bath. Figure 3 is a

Table 7
BEER-LAMBERT CONSTANTS FOR AROMATIC HYDROCARBONS IN CYCLOHEXANE

	λ max, nm	Range, mg/ℓ	Beer-Lambert, constant 1 mg^{-1} cm^{-1} 10^3 std dev
p-Dichlorobenzene	273	90—294	0.00378 ± 0.01
Naphthalene	277	3.2—7.2	0.0455 ± 0.2
Acenaphthene	289	3.2—7.2	0.0433 ± 0.1
Biphenyl	248	6.2—42	0.1094 ± 0.3
Fluorene	301	2.1—8.3	0.0602 ± 0.4
Phenanthrene	293	0.6—5.0	0.0832 ± 0.3
Anthracene	253	0.3—1.3	1.294 ± 5
Pyrene	335	2.0—8.3	0.278 ± 1

Reprinted with permission from Wauchope, R. D. and Getzen, F. W., *J. Chem. Eng. Data,* 17, 38, 1972. Copyright by the American Chemical Society.

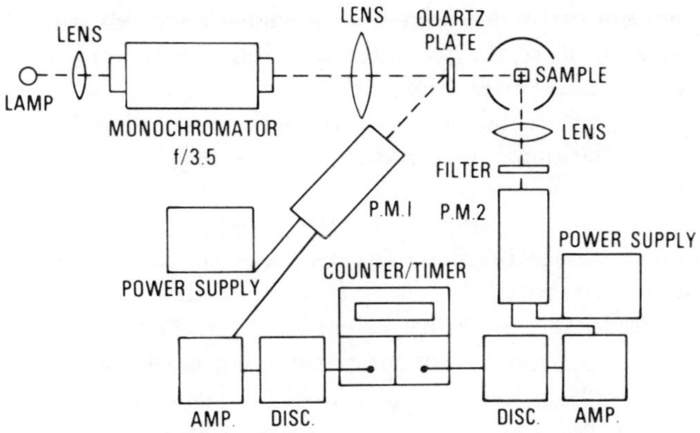

FIGURE 3. Schematic of the fluorescence apparatus M_1 is the f/3.5 monochromator, P.M. is the excitation photon measuring photomultiplier and P.M. 2 is the fluorescence photon measuring photomultiplier.

schematic of the fluorescence apparatus used and it is described in Reference 27. Some of the key features of this apparatus include a 60 W deuterium excitation source and a ratio-photon counting detector designed to minimize the effects of light source intensity variations. At fluorescence absorptivities of less than 0.05, the concentration C of PAH is related linearly to the fluorescence signal F,

$$F = k_1 C + k_2 \tag{4}$$

where k_1 is a constant at a fixed excitation and emission wavelength and depends on a number of factors such as instrumental variation, the constant pre-set excitation photon count, fluorescence quantum yield, cell depth, magnitude of oxygen quenching, and temperature. The intercept, k_2, is a constant composed mainly of contributions due to thermal noise of the photomultiplier and is constant over a wavelength range of 300 to 425 nm. For each determination of concentration, k_2 is determined at a wavelength where A = O, i.e., no fluorescence. Some of the solubility results are listed in Table 2, where in each case fluorescence measurements were performed upon the previously determined optimum wavelength of excitation and emission.

In addition, McKay et al.[17] have used a fluorescence method employing an Aminco-Bowman® spectrophotofluorometer. Instrumental details are not presented. By means of standard calibration curves of PAH in cyclohexane, the fluorescence intensity is found to be linearly related to concentration from 1 μg/ℓ. up to 1000 μg/ℓ. Reproducibility is in the 5 to 10% range with the precision higher for the more soluble PAH.

In the DCCLC technique, the PAH solution from the analytical column is pumped through an integrated UV detector where the signal is proportional to the concentration. To perform a quantitative analysis, standard PAH solutions in acetonitrile are tested to determine individual response factors. The sample loop volume is determined by an indirect gravimetric procedure before studies are carried out on the aqueous solubilities of PAH. The solubility results are also presented in Table 2. The sources of error for the DCCLC technique in the quantitative steps include the calibration of the sample loop volume, preparation of the standard PAH solutions in acetonitrile, and the volumetric measurements of the amount of sample per analysis. Quadratic addition of the random errors associated with each step yields a total error of 1.6% for the quantitative analysis of the generated solutions. Long- and short-term precision for the DCCLC technique is better than ±3%.[23]

The difficulties inherent in the fluorescence method include such factors as oxygen quenching, detector variations, and nonanalyte interferences. A precleanup step before analysis by fluorescence might help resolve the latter problem.

An overall examination of Table 2 indicates quite good reproducibilities between different labs and different techniques. The results of Davis, determined some 35 years ago by nephelometry are reasonably close to modern day values. A much discussed discrepancy[20,24] exists in the solubility values listed in Table 2 for anthracene. The values of Schwarz, and May et al. are comparable but are lower by nearly a factor of two from the other results reported. Some reasons suggested for the difference are (1) impurities of phenanthrene in the anthracene causing an increase in apparent anthracene solubility due to the 20-fold greater aqueous solubility of phenanthrene, (2) the use of a dilute acetone water solvent in the nephelometric approach[24] thereby raising the apparent solubility of anthracene, and (3) an extrapolation of the anthracene solubility at 25°C in Reference 18 from the solubility above 35°C resulting in a high positive error. A fact left unexplained, though, is that McKay's value is also high using fluorescence detection presumably similar to Schwarz.

Temperature dependencies of the solubilities were determined by UV,[18] fluorescence,[17,20] and DCCLC with UV detection.[24] The results of May et al.[24] are graphically illustrated in Figure 4.

Other solubility parameters which were investigated and are of some interest include Setschenow constants which can be used to predict solubilities of PAH in salt solutions, i.e., ocean waters. The results of these studies are discussed in a number of references.[20,21,24]

III. APPLICATION OF SOLUBILITY DATA TO ENVIRONMENTAL SYSTEMS

It is of great analytical interest to be able to measure the aqueous solubilities of

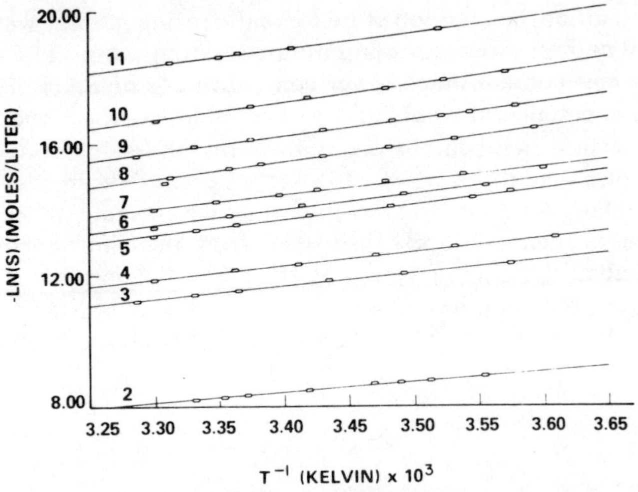

FIGURE 4. Dependencies of the solubility of several PAH on temperature. Key to Figure 4 2. Naphthalene 3. Fluorene 4. Phenanthrene 5. 1-methylphenanthrene 6. Pyrene 7. Fluoranthene 8. Anthracene 9. 2-methylphenanthrene 10. Benzo(*a*)anthracene 11. Chrysene.

PAH to the best possible accuracy and precision. However, a number of factors make the precise determination of solubilities in environmental waters a near impossibility. First, in nature it is not likely that isolated PAH will be found. More likely numerous PAH, unsaturated hydrocarbons, and other organics and inorganics will be present. A quantitative analysis of all these combined effects can only be approximate at best.

It has been noted that various impurities in waste waters could increase PAH solubilities.[28] Phenol concentrations greater than 50 mg/ℓ increased the solubility of B(a)P, whereas lesser amounts of phenol had no effect. In detailed laboratory studies on finding a suitable laboratory solvent for B(a)P,[29] it was also found that solubility increased with increasing organic solvent concentrations. Therefore, detergency effects in natural water systems and the existence of other organics can alter the water solubilities of PAH. It is important that in areas where waste waters might contain high levels of organic impurities that this consideration be made.

Another problem with environmental systems is that of evaporation. Results on PCBs, naphthalene, and biphenyl as well as lower molecular weight hydrocarbons indicate that significant amounts of hydrocarbon can be lost from water by evaporation,[30] with the greater losses occurring for the more volatile species. Even higher molecular weight PAH can be lost from water by evaporation due to the extremely high activity coefficients of these compounds in water causing high equilibrium partial pressures and consequently high rates of evaporation. This effect can create a difficulty in determining solubilities of PAH in natural waters.

Finally, studies in water on the partitioning of anthracene between a dissolved and adsorbed state on organic yeast particles indicate a large fraction of anthracene in the adsorbed state.[31] A varying amount, 15 to 65%, of anthracene was found to be adsorbed, with the higher percentage occurring at high levels of yeast particles. Many waste waters can contain high levels of organic particulate matter, and thus a significant fraction of PAH may be found adsorbed and not in solution. The relationship between adsorption and solubility is a factor which cannot be ignored for real-life samples.

In addition to all of the mentioned factors influencing natural water solubility, the assumption that natural waters contain saturated concentrations of PAH is rarely true. Most water systems contain much lower concentrations of PAH. The ability to precisely correlate experimental solubilities with solubilities in aqueous environments is not possible, but the experimentally determined solubilities do define a reasonable upper concentration range to be expected in the absence of detergency type effects. Instead, experimental solubilities do offer important insights into understanding solution theory more fully. Their usefulness also stems from their introduction of potentially more useful analytical techniques such as the DCCLC approach. The empirical approaches towards predicting solubilities are quite valuable for environmental applications since their degree of uncertainty does not seriously affect the sought after results.

Chapter 3

SAMPLING AND PRECONCENTRATION

I. INTRODUCTION

The determination of polycyclic aromatic hydrocarbons (PAH) in water systems is at present being studied with great interest since a large number of PAH have been shown to be high carcinogenic. The analyst is confronted by a number of problems in the analysis of PAH in water samples.

The first problem is that the concentration of individual PAH range from less than 1 ppt (pg/g) in pure groundwater supplies to greater than 1 ppm (μg/g) in heavily contaminated sewage. These concentrations necessitate the application of some extraction or preconcentration technique to raise the concentrations to a level at which identifications can be made and quantitative analyses carried out.

The second problem occurs when handling solutions where the concentrations of the solute is in a range less than 1 ppm. Serious errors can arise from losses or contamination in sampling or indeed in any step of the analytical process. Practically any surface with which the sample comes in contact can become a source of contamination, or a surface on which the trace pollutant can be lost by adsorption, reaction, etc.

The third problem is that the PAH may represent as little as 0.01% of the organic fraction present in the water sample, thus, the analytical scheme must be devised so that the PAH can be analyzed without interference from other pollutants.

The literature relating to the extraction and preconcentration of PAH in water has been carefully evaluated. Many of the studies reported in the literature have been carried out with model solutions of PAH, thus necessitating careful consideration before applying the results to an environmental ('real') sample which may contain a large number of other organics at concentrations which may mask the PAH present in the sample. Some of the procedures are accurate and relatively simple, enabling a large number of samples to be rapidly and inexpensively treated by nonprofessional personnel. Other collection and concentration methods are examples of superb analytical techniques, but are very time consuming and involve complex equipment that require the skills of a very highly trained chemist.

The techniques, reviewed here, that have been used to extract and concentrate the PAH from water samples include adsorption methods, solvent extraction, headspace analysis, steam-distillation, and coupled-column chromatography.

II. COLLECTION AND STORAGE OF WATER SAMPLES

Two approaches are available to the analyst for the collection and separation of the PAH. Samples can be collected in the field and returned to the laboratory for working-up, or they can be preconcentrated on-site. In the cases where the samples are stored prior to the extraction of the PAH, they must be stored in glass containers with metal liners in the plastic cap. Furthermore, they should be stored in the dark at subzero temperatures (°C) to minimize reactions leading to the loss of PAH.

Several groups of investigators have studied this problem quantitatively. Hertz et al.,[32] working with stirred solutions of PAH at the 1 ppb level, showed that adsorption losses of 80% occurred in glass containers after 4 hr of stirring and that these losses increased to 95% after 40 hr of stirring. Hertz et al. also observed that the freeze-drying of samples for preservation resulted in the loss of volatile components. Strup et al.[33] obtained low recoveries from water samples spiked in the ppb range and showed that these low recoveries were due to adsorption losses on the walls of the glass container used in their experimental setup. Recovery factors were 50% higher when the PAH were directly spiked into the sampling module rather than when the PAH first passed through the water reservoir being studied before entering the sampling module. Basu and Saxena[34] have followed the losses for six PAH used in the preparation of 60-ℓ standard water solutions where the PAH concentrations were in the ppb range. The adsorption losses ranged from 44.3% in the case of fluoranthene to 77.4% for benzo(*ghi*)perylene and are listed in Table 8. Acheson et al.[35] also studied the effect of a 6-hr mixing on the recovery of PAH from water solutions both with and without added solids. The added solids did not appear to affect the extraction efficiency. They concluded that losses were due either to the adsorption of PAH on the glass of the mixing and extraction vessels or to chemical or biological degradation within the vessels.

Glassware used in PAH analyses is generally cleaned by either detergents or organic solvents or both. Some workers have followed this treatment by heating the glassware to 400°C to ensure the removal of traces of organic matter. Ogan et al.[88] have noted that silanization of glassware and flushing of all components of the system with the sample improved recoveries. The desirability of further studies with a view to the development of a standard procedure for the treatment of sampling apparatus is indicated.

The important conclusion from all of these observations is that if possible one should sample directly into the extraction vessel and carry out the extraction as soon as possible after sample collection.

III. SOLID ADSORBENTS

A. Tenax-GC®

Tenax-GC® is a porous polymer based on 2,6-diphenyl-p-phenylene oxide. It has found use as a packing material in gc work and has been used for the collection of PAH from water and air samples.[36] The outstanding advantages in using Tenax-GC® are that (1) the adsorption and recovery of organics from the polymer are almost complete, (2) the polymer is impervious to water and, (3) the polymer has excellent thermal stability.

Leoni et al.[37] have studied Tenax-GC® for the extraction of PAH and pesticides from surface and drinking waters. They examined a number of parameters of the sampling system such as (1) the amount of Tenax-GC® in the sampling column, (2) the effect of pH, (3) the flow rate of the water sampled on the collection efficiency, and (4) the volume of solvent required to elute the hydrocarbons from the Tenax-GC.® On the basis of their studies they selected a column 7.5 cm × 1 cm containing 1.5 g of 60 to 80 mesh Tenax-GC.® The column was used to sample up to 20 ℓ of water at the rate of 3 ℓ/hr. The optimum pH of the water was between 6.8 to 7.2. Under these conditions, analysis of the column effluent showed no traces of PAH or pesticides with the exception of small amounts of hexachlorobenzene. The adsorbed pollutants were extracted from the Tenax-GC® with 30 mℓ of acetone. Leoni et al. give results for the addition of five different PAH to 25 to 30 ℓ of drinking water at levels of 0.08 to 0.13 ppb and report recoveries that average 95%. They note that the adsorption of pollutants by polymers can be adversely affected by the presence of surfactants and fatty substances and recommend the use of a Celite® 545 prefilter when working with natural waters. We note that the removal of interfering substances such as particulate matter, fats, etc., poses serious problems since as we discuss below in the presence of small particulate matter about 50% of the PAH may be lost from the water by adsorption.

Table 8
AMOUNT OF PAH UNACCOUNTED FOR AS A RESULT OF MIXING WITH WATER IN A GLASS BOTTLE

Compound	Concn. on basis of amt. added to water (ng/ℓ)	Concn. found in aq. phase (ng/ℓ)	% PAH losses
Fluoranthene	500	278.6	44.3
Benzo(*j*)fluoranthene	100	48.3	51.7
Benzo(*k*)fluoranthene	100	51.7	48.3
Benzo(*a*)pyrene	100	36.4	63.6
Indeno(*1,2,3-cd*)pyrene	100	25.5	74.5
Benzo(*ghi*)perylene	100	22.6	77.4

Note: Water volume: 60 ℓ.

B. Amberlite® Resins

The Amberlite® resins have received a great deal of attention as adsorbents of organic species in water. The XAD 1-5 resins are polystyrenedivinylbenzene copolymers and are nonpolar; whereas XAD-7 and XAD-8 are polar polymethacrylate resins.[38,39]

Prior to their use in the collection and concentration of organic compounds from water systems the Amberlite® resins must be cleaned to remove impurities or nonpolymerized material that will be extracted when the water pollutants are desorbed from the resin. All investigators who have used Amberlite® resins to extract pollutants from water systems have used essentially the same cleanup procedure. Stepan and Smith[38] have used a 6-hr Soxhlet extraction with methanol followed by a 6-hr Soxhlet extraction with diethyl ether. After the second extraction the resin is equilibrated with 20 mℓ of diethyl ether for 10 min and the diethyl ether eluate is analyzed by gc to check the effectiveness of the cleanup. If the blank chromatogram indicates that the resin is clean, the resin is stored under methanol in a glass bottle until it is to be used. Storage of the resin under methanol prevents adsorption of organic pollutants from the atmosphere and also prevents the resin from drying out. It has been shown that the dried resin develops cracks allowing more impurities to leach out of the resin.[40] Fritz et al. who have used Amberlite® resins in a number of studies of water systems use essentially this same cleanup procedure.[40-44]

Strup et al.[33] have carefully studied the resin blank problems associated with the use of XAD-2 resins. In their cleanup procedure the resins were cleaned with five to six washes of water followed by 24-hr Soxhlet extractions with methanol, diethyl ether, pentane, and methylene chloride. One batch of the cleaned resin was dried and stored under nitrogen; a second batch was washed with methanol and stored as a methanol slurry. Each of these resin samples was used to analyze 4-ℓ portions of the same batch of distilled water. The adsorbed solutes were eluted from the resins by an 8-hr Soxhlet extraction with methylene chloride and the volume reduced in a Kuderna-Danish evaporation apparatus. The gc traces, Figure 5 and 6, show the superiority of the slurry storage method over the dry storage method. Our own studies with Amberlite® XAD-2 and XAD-4 also indicate that sequential 24-hr Soxhlet extractions with methanol, acetronitrile, and diethyl ether with subsequent storage under methanol is necessary in order to obtain a satisfactory gc blank.

The selection of a particular XAD resin for the extraction of PAH from water sys-

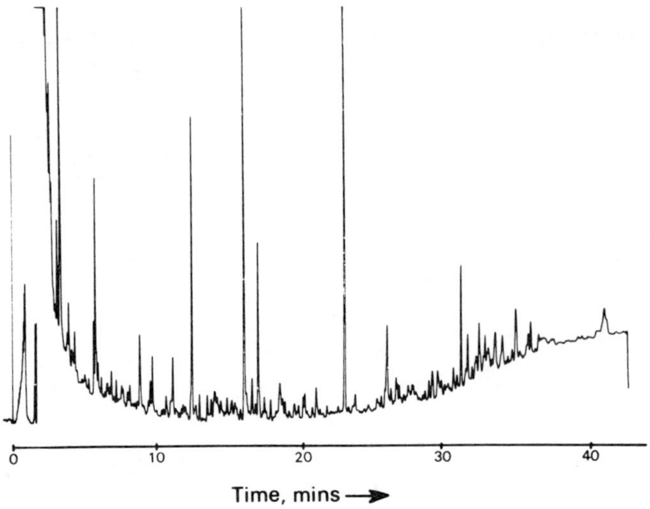

FIGURE 5. Gas chromatographic scan of the eluent from an XAD-2 blank, after dry storage and dry packing. (From Strup, P. E., Wilkinson, J. E., and Jones, P. W., *Polynuclear Aromatic Hydrocarbons,* Raven Press, New York, 1978. With permission.)

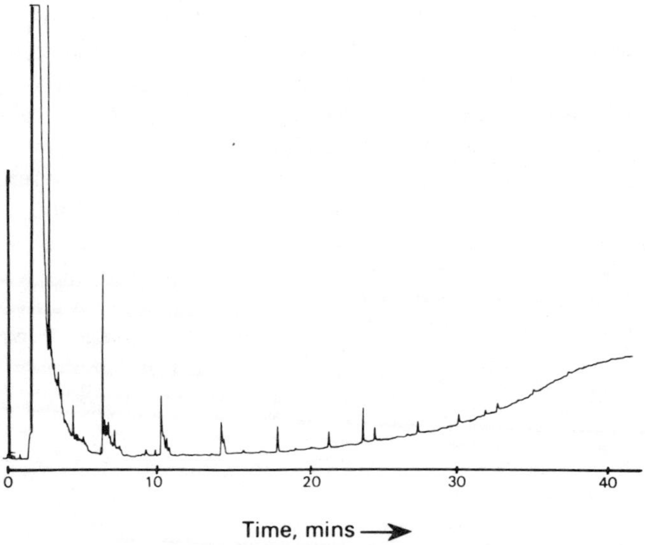

FIGURE 6. Gas chromatographic scan of the same XAD-2 blank used in Figure 5, but with methanol storage and slurry packing. All other conditions and chromatographic parameters were the same as in Figure 5. (From Strup, P. E., Wilkinson, J. E., and Jones, P. W., *Polynuclear Aromatic Hydrocarbons,* Raven Press, New York, 1978. With permission.)

tems appears to be determined by the personal bias of the investigator. Fritz et al. have used both XAD-2 and XAD-4 at various times.[41,45] The recovery efficiencies of the XAD-2, 4, 7, and 8 resins and resin mixtures have been studied by Van Rossum and Webb.[46] In their comparative study they used distilled water samples spiked with 13 organic pollutants. For tap water samples they found XAD-2 and an equal-weight mixture of XAD-4 and XAD-8 to be about equally effective, The average recovery was about 80% for acenaphthene, acenaphthylene, fluoranthene, and pyrene. However, for extracting chlorinated insecticides and polychlorinated biphenyls from water, Musty and Nickless[47] have shown that XAD-4 is far superior to XAD-2.

Thurman et al.[48] have studied the capacity factors for organic solutes in water systems when adsorbed on Amberlite® XAD-8 resin. Their results can be used as a basis for estimating the size of the resin column required to sample PAH from water. Factors in addition to the resin type can affect the efficiency with which organic pollutants are extracted from water systems. Stepan and Smith[38] have shown that the recovery factor of naphthalene decreases when the rate of flow of water through the adsorption column is increased from 20 to 200 mℓ/min. The extraction efficiency of the column for nonpolar compounds such as PAH are independent of the pH of the water, however, the extraction of organic acids is strongly pH dependent.[41] Junk et al.[41] have investigated the effect of varying the particle size of the resin. They tested spherical resin beads, 20 to 60 mesh, as well as resin samples that had been ground to meshes as small as 150. No advantage to any particular size range was found and they recommend the 20 to 60 mesh spherical resin.

For sampling, Stepan and Smith[38] filled the glass adsorption tube with a slurry of Amberlite® in methanol and plugged the ends of the tube with either sintered glass or glass wool. Thirty milliliters of methanol were passed through the adsorption tube and before the methanol had completely drained, the tube was stoppered with glass stoppers until it was to be used. Immediately before sampling, two 100-mℓ portions of distilled water were passed through the adsorption tube to remove the methanol.

For the study of model samples in the laboratory and the analysis of grab samples collected in the field, Junk et al.[41,42] used the apparatus shown in Figure 7. A silanized glass wool plug (E) is inserted near the stopcock. The methanol slurry of XAD resin (F) is added to form a bed approximately 6 cm high containing 1.5 to 2 g of dry resin. A second plug of silanized glass wool (E) is placed over the resin, the methanol is drained, and the resin is washed with three 20-mℓ portions of water. During this step, care is taken to drain the wash liquids only to the top of the resin bed. The reservoir flask(C) of the apparatus shown in Figure 7 has a capacity of either 2 or 5 ℓ. The water sample is placed in the reservoir and the top of the reservoir is connected to a cylinder containing organic-free nitrogen. The water sample is passed through the resin bed at a rate of 30 to 50 mℓ/min. If the water flow falls below this rate, nitrogen at about 1 psi is applied to the reservoir to increase the flow rate. When most of the sample has drained through the resin, the walls of the reservoir are washed with two 20-mℓ portions of pure water, and the liquid level is allowed to drop below the level of the resin only after the last wash. After the water has drained from the column the sample is extracted with diethyl ether. First the reservoir is washed with two 10-mℓ portions of diethyl ether which are allowed to drain into the column and are retained there for about 10 min. During this period, the reservoir is removed and the column is capped with a 24/40 glass stopper. After the 10-min equilibration period, the ether is drained into a test tube and an additional 5 mℓ of diethyl ether is passed through the column.

Immediately after the sample extraction, the column is regenerated by passing 30 mℓ of methanol through the column and shaking the column to remove entrapped air bubbles. The stopcock is then closed, 15 mℓ of methanol is added, and the column is ready for use again without any additional preparation.

For composite sampling over a 24-hr period, Junk et al.[43,44] used the column shown

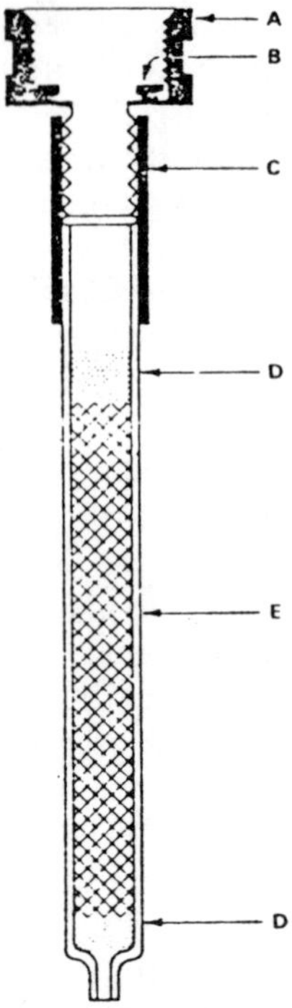

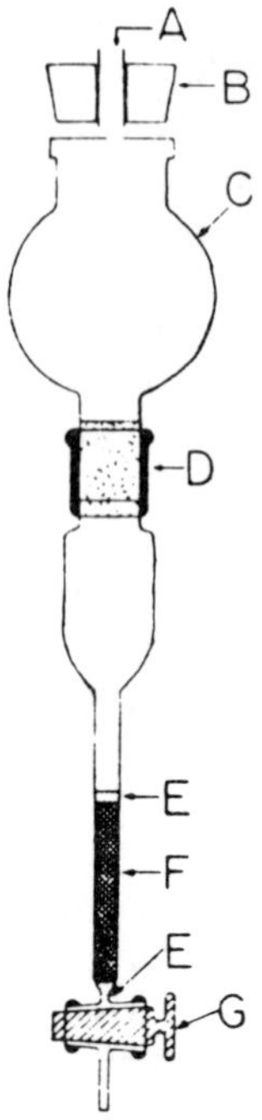

FIGURE 7. Apparatus for extracting organic solutes from water. (A) purified inert gas pressure source; (B) cap; (C) 2-ℓ reservoir; (D) 24/40 ground glass joint; (E) silanized glass wool plugs; (F) 0.6 cm-i.d. × 10 cm-long glass tube packed with 20 to 60 mesh XAD-2 resin; (G) PTFE stopcock. (From Junk, G. A., Richard, J. J., Grieser, M. D., Witiak, D., Wihak, J. L., Arguello, M. D., Vick, R., Svec, H. J., Fritz, J. S., and Calder, G. V., *J. Chromatogr.*, 99, 745, 1974. With permission.)

FIGURE 8. Illustration of composite sample extraction device, (A) standard garden hose coupling; (B) PTFE washer; (C) 12.7 mm-i.d. PTFE tubing; (D) glass wool plugs; (E) 12.7 mm-o.d. × 9 cm-long glass tube packed with approximately 2 g 40 to 60 mesh XAD-2 resin. (From Junk, G. A., Richard J. J., Svec, H. J., and Fritz, J. S., *J. Am. Water Works Assoc.*, 68, 218, 1976. With permission.)

in Figure 8. The glass tube was filled with XAD-2 resin as described for the column in Figure 7 and connected to a standard garden hose coupling with a short length of PTFE tubing. The water is sampled through this extractor at the rate of 150 mℓ/min. so that over a 24-hr period 200 ℓ of water is sampled. After the sampling, the garden hose coupling is replaced by a 24-mℓ bulb, retaining the same piece of PTFE, and a PTFE stopcock is attached to the bottom of the column with another piece of PTFE tubing. The PAH are eluted with diethyl ether as described above.

Strup et al.[33] have modified a commercially available 400 × 25 mm liquid chromatography column (Fischer Porter Co.) to serve as a composite sampling tube. The water sample is drawn through the tube with either an aspirator or a pump.

After the adsorbed sample has been eluted from the adsorption column, the 20 to 30 mℓ of eluate is generally reduced in volume to about one mℓ. To ensure both a smooth volume reduction and to avoid the introduction of water into the gc, the eluted sample is first dried. Stepan and Smith[38] used magnesium sulfate as the drying agent. Junk et al.[41] have investigated a number of alternatives for removing the water from the eluate obtained from the column. They stated that placing the eluates in a refrigerator for a 24-hr period introduced evaporative and irreversible adsorption losses. They also noted that although rapid cooling with ice-salt or dry-ice-acetone slush baths did not result in losses of PAH, the handling and preparation of these baths is inconvenient. They found that a sodium sulfate drying procedure or a liquid nitrogen freeze-out was completely satisfactory. The liquid nitrogen freeze-out for two 10-sec intervals is the recommended technique since it avoids the addition of an extraneous component to the eluate which might contaminate the solution or serve as a surface on which the solute can be adsorbed.

For samples in which the PAH concentration in the eluate is too low for reliable determination, volume reduction is necessary prior to gc. Our experience and the experience of other workers[40] has shown that the concentration step is a critical step where serious solute losses can occur. Junk et al.[41] recommend a distillation technique to concentrate the sample. They have used the apparatus shown in Figure 9 with a Snyder column when the sample volume is to be reduced to 1 mℓ and a Vigreaux column when the sample is to be reduced to 0.6 mℓ. The Snyder column is less likely to lead to losses when bumping occurs; on the other hand the Vigreaux column has negligible holdup. They use a boiling chip and an evaporation rate of between 0.5 to 2.0 mℓ/min. They state that small boiling chips do not adsorb organic solutes. Smith et al.[49] have used a rotary evaporator for concentrating solutions of PAH. They found that with a little experience the concentration proceeds smoothly.

Junk et al.[41] examined a number of pretreatment techniques to test glassware activity. They tested (1) sequential solvent washing with acetone, methanol, and ether, (2) detergent washing followed by the organic solvent washings, (3) detergent washing and drying, (4) silanization of the glass surfaces, (5) baking the glassware at 400°C for 16 hr, and (6) substitution of Clear-Seal® for ground glass joints. They were unable to correlate any of these treatments with surface activity. For this reason they settled on and recommend sequential solvent washing for all glassware.

Junk et al.[41] have also investigated the shape of the vessel in which the concentration step is carried out. They have studied a number of configurations of concentration vessels. Several of the shapes that were studied are shown in Figure 10. When vessels B and C were used solute losses from 10 to 60% were noted, whereas losses from vessels of shape A were at the most 6%. The vessels of shape A are constructed by sealing a graduated centrifuge tube to a round bottom ground glass joint flask, taking care to produce a smooth seal. This configuration gives an effective solution washing of the sides of the flask during the solute reduction step.

Junk et al.[41] have also investigated the volume reduction by using a stream of nitrogen or another inert gas. Their results show that this method leads to serious solute losses and should be avoided. They point out that 10 to 80% of all solutes except those of extremely low volatility are lost by using a gas stream to concentrate the sample. Shinohara et al.[50] have compared the recovery factors for the nitrogen gas stream method and the micro-Snyder distillation method for volume reduction for benzene and ether solutions of PAH. They found that the average recovery from the distillation method was about 5 to 7% higher than that for the nitrogen gas stream method.

Both investigations show that the micro-Snyder distillation method is superior to the nitrogen gas stream method, but losses by the nitrogen gas stream method are dependent on the individual investigator. The problem of reducing the volume of the sample to a predetermined final volume (1 mℓ), is not a simple one. Junk et al.[41] suggest that in order to arrive at a desired final volume, the volume of the solution be reduced below this level and then brought up to volume by the addition of ether. Smith et al.[49] have used this same approach but they adjusted the final volume with isooctane.

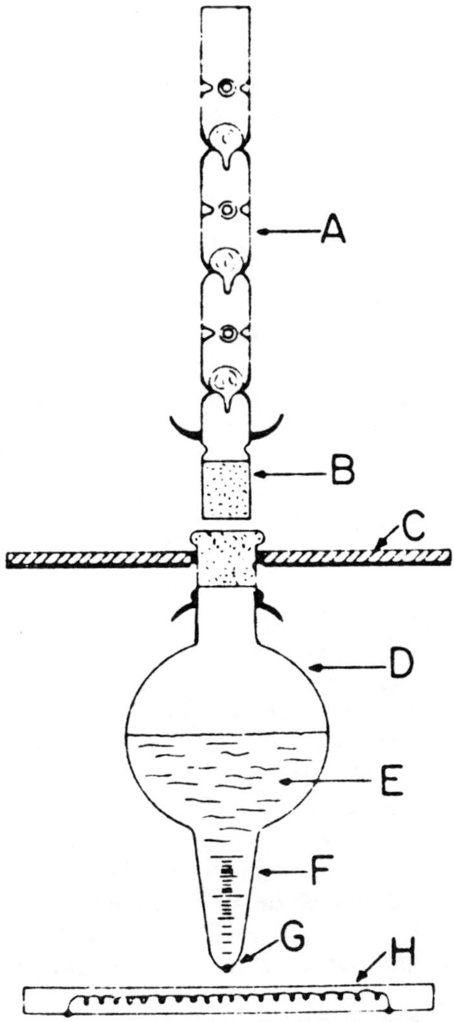

FIGURE 9. Sketch of a concentration apparatus. (A) Snyder distillation column; (B) 14/20 ground glass connection; (C) Bakelite® heat shield covered with Al foil; (D) approximately 50-mℓ vessel; (E) ether solution; (F) graduated and calibrated taper; (G) small boiling chip; (H) hot plate. (From Junk, G. A., Richard, J. J., Grieser, M. D., Witiak, D., Wihak, J. L., Arguello, M. D., Vick, R., Svec, H. J., Fritz, J. S., and Calder, G. V., *J. Chromatogr.*, 99, 745, 1974. With permission.)

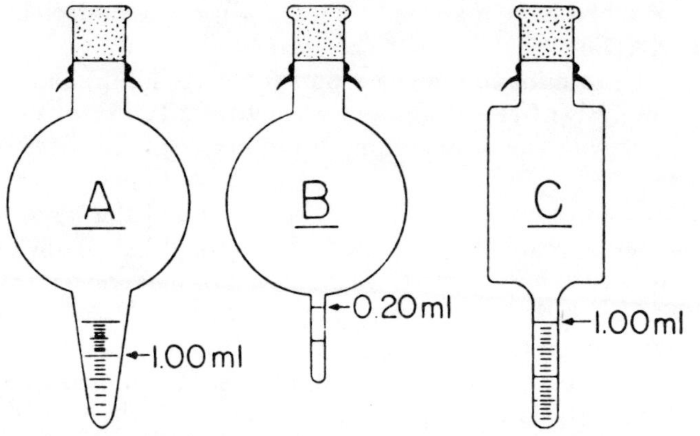

FIGURE 10. Various configurations of concentration vessels tested. (A) is recommended, (B) is unsatisfactory, (C) is questionable. (From Junk, G. A., Richard, J. J., Grieser, M. D., Witiak, D., Wihak, J. L., Arguello, M. D., Vick, R., Svec, H. J., Fritz, J. S., and Calder, G. V., *J. Chromatogr.*, 99, 745, 1974. With permission.)

The higher boiling point of the isooctane reduced solvent losses during the handling of the sample.

Tateda and Fritz[45] have recently reported a procedure that uses a mini-column requiring only a small water sample. In this way, the volume reduction step is eliminated. The mini-column, constructed of a disposable pipet, was packed with an adsorbent bed 1.2 to 1.8 mm × 25 mm consisting of XAD-4 or Spherocarb® (Analabs, New Haven, Ct.). By using this column, the volume of water sampled could be reduced from 2 to 5 ℓ to 50 to 100 mℓ and the volume of eluent reduced to 50 to 100 μℓ. Thus, without resorting to a volume reduction, the concentration of the PAH in the mini-column eluate is 10 to 20 times greater than that obtained after the eluate from the larger column has been reduced from 25 mℓ to 1 mℓ. However with the smaller volume of adsorbent, the sensitivity of the mini-column method is limited to 2 ppb for each organic solute in the water.

Sampling with the mini-column is carried out in the same way as with the larger columns, but at a flow rate between 0.8 to 1.2 mℓ/min. Since the volume reduction step is eliminated in the XAD-4 mini-column method, the more volatile diethyl ether that is used to elute the sample from the larger column can be replaced by acetone. Fifteen to twenty microliters of acetone are added at a time and forced back and forth through the resin bed with a rubber bulb until a total of 50 μℓ of acetone has been added. For low molecular weight and water soluble compounds the elution is carried out with carbon disulfide added in 20 μℓ portions. For the mini-column the average recovery was 89% at the 2 ppb level and 83% at the 100 ppb level; this compares favorably with the average recovery of 92% for the same compounds with the larger apparatus.

The same model compounds that were used to test the mini-column packed with XAD-4 were used to test the mini-column packed with Spherocarb®. Elution of the solutes from the Spherocarb® column was both slow and difficult. Carbon disulfide was the best eluent, superior to acetone, methanol, diethyl ether, acetonitrile, methylene chloride, and pentane; but even with carbon disulfide a number of compounds were difficult to elute. Five successive 100-μℓ portions of carbon disulfide failed to completely elute naphthalene, biphenyl, and 2-naphthol from Spherocarb.® Thus, the

Spherocarb® appears to have a stronger retention for low molecular weight organic compounds than XAD-4.

The sensitivity of the mini-column is 2 ppb for each solute and this sensitivity is adequate for the analysis of raw water and wastewater. The results of the analysis of 100 mℓ of raw water with the mini-column using both Spherocarb® and XAD-4 are shown in Figure 11.

Tateda and Fritz[45] have combined the mini-column and the large scale adsorption method to take advantage of the best features of both. Whereas the large scale apparatus has good sensitivity, only a small fraction of the final concentrate (2 μℓ out of 1 mℓ) is used for the gc analysis. The mini-column does not have adequate sensitivity for the determination of pollutants in pure water, but the eluate from the column is 50 to 100 μℓ. The best features of both methods were combined by taking the 1 mℓ ether concentrate from the larger column, adding it to 50 mℓ of pure water and processing this sample in the mini-column. When 2 μℓ of the 100 μℓ carbon disulfide extract from the mini-column is analyzed a further improvement by a factor of 10 in sensitivity is achieved. The enhancement in sensitivity resulting from the second concentration by the mini-column is shown in Figure 12.

Malcolm et al.[51,52] have quantitatively related the separation of organic solutes on XAD resins into hydrophobic and hydrophilic fractions. Their papers deal with the hydrophobic fraction which contains the PAH compounds. Although they have not tested PAH compounds, the classification is pertinent to this review. They have selected XAD-8 for their study, having found it the most effective adsorbent for both high and low molecular weight organic solutes in natural waters. XAD-8 was chosen in preference to XAD-2, since the latter resin shows a tendency to become clogged by natural polyelectrolytes and this results in a reduced retentive capacity of the adsorption column.

The major factor that influences the retention of organic solutes on XAD-8 is the solvent-solute interaction rather than the solute-resin interaction. Solvent-solute interaction can be related to the formation or disruption of the hydrogen-bonded cage structure of the water molecules and can be expressed quantitatively as the solubility of the organic solutes in water. Thurman et al.[48] have correlated the logarithm of the column distribution coefficient, k^1, with the logarithm of the solubility as shown in Figure 13. With a knowledge of the solubility, the retention of an organic solute can be predicted to within 10%. The column distribution can be defined as:

$$k^1 = \frac{\text{mass of solute in stationary phase}}{\text{mass of solute in mobile phase}} \tag{5}$$

From the graph it can be seen that for a homologous series the amount of material adsorbed by the resin column increases as the molecular weight increases or as the aqueous solubility decreases.[39,53] Since the molecular weight of PAH are generally higher and their aqueous solubilities lower than most other organic solutes, the Amberlite® resins will serve as an excellent matrix for the adsorption of PAH from water.

Leenheer and Huffman[54] have proposed the use of macroreticular resins as both a classification and adsorption tool. The macroreticular resins are used to concentrate and separate the dissolved organic carbon into a hydrophobic and hydrophilic fraction. The hydrophobic/hydrophilic break is defined for a $k^1 = 100$ on XAD-8. In the analytical DOC fractionation,[55] the hydrophobic solutes are adsorbed on a nonionic resin of low polarity (XAD-8) and the hydrophilic fraction is collected on macroreticular ion-exchange resins. The solutes are subsequently desorbed from the resins and separated into acidic, basic, and neutral fractions. The technique merits attention since it

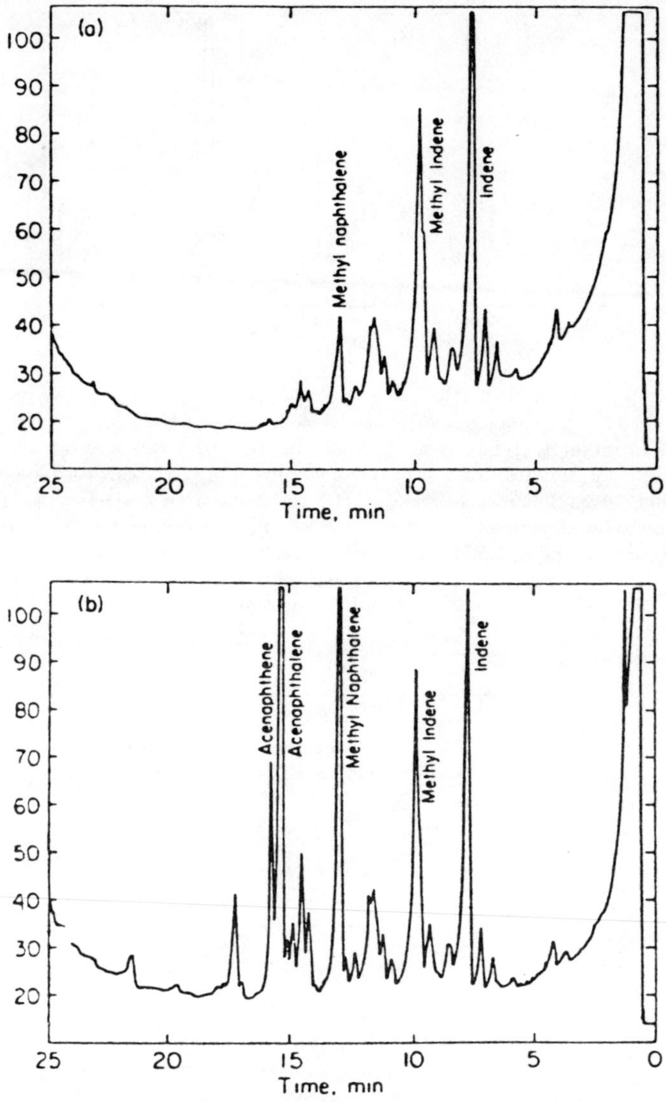

FIGURE 11. Chromatograms of well water in Ames, Iowa collected on Spherocarb® and XAD-4. (A) Eluate from Spherocarb®; (b) eluate from XAD-4. Gas Chromatographic column, 1/8 in. × 6 ft., 3% OV-17; temperature, 30-240° at 8°/min; temperature of injector and FID detector, 350°; carrier gas flow-rate, 20 mℓ/min. (From Tateda, A. and Fritz, J. S., *J. Chromatogr.*, 152, 329, 1978. With permission.)

bridges the gap between the analysis of total dissolved organic carbon and the specific compound identification of all organic species dissolved in water that one would obtain with a gc/ms system. The technique is presented by Leenheer and Huffman as a total classification scheme for DOC. Of special interest in this case is that the procedure can be used to separate the PAH (which are a part of the neutral hydrophobic fraction) from the acidic and basic hydrophobic fractions and the hydrophilic compounds.

Leenheer and Huffman[54] define hydrophobic solutes as the organic compounds that are adsorbed on resin surfaces of low and intermediate polarity. The polarity require-

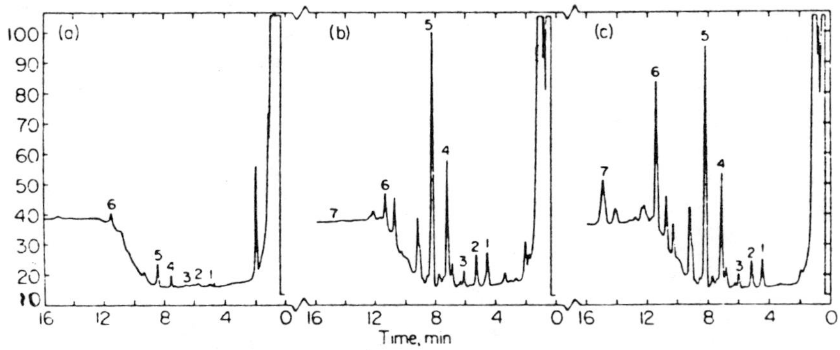

FIGURE 12. Chromatograms of Iowa State University tap water after concentration by (I) large-scale sorption, and (II) large-scale sorption followed by mini-column sorption (see text). (a) Ether concentrate by large-scale sorption method, (I) (b) Eluate from Spherocarb® method (II); (c) Eluate from XAD-4 method (ii). Gas chromatographic column 6 ft. × 1/8 in., 10% FFAP; temperature, 60-200° at 16°/min; initial temperature hold, 2 min; final temperature hold, 8 min; temperature of injector and FID detector, 300°; carrier gas flow-rate, 20 mℓ/min. (From Tateda, A. and Fritz, J. S., *J. Chromatogr.*, 152, 329, 1978. With permission.)

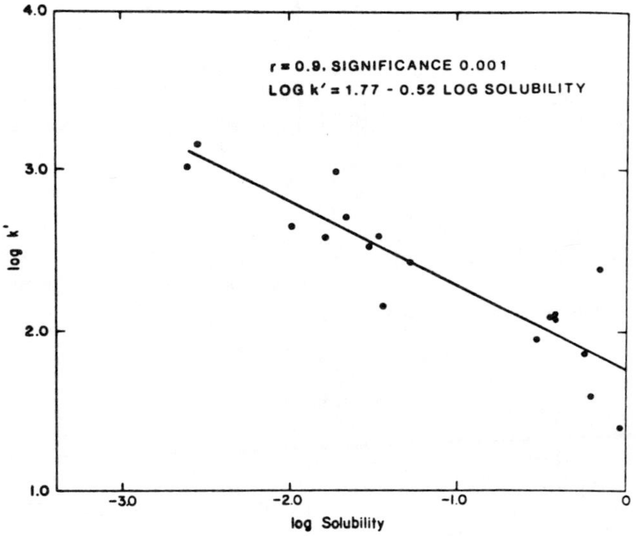

FIGURE 13. Log k′ vs. log solubility for 18 organic solutes.

ment is met by XAD-2, 4, and 8 resins. The hydrophobic solutes repel water and are attracted to the hydrocarbon surface of the resin by van der Waal's forces. The hydrophilic solutes have water-attracting oxygen, nitrogen, or sulfur functional groups such that these solutes interact with water through hydrogen bonding.

In the "analytical DOC fractionation" procedure outlined by Leenheer and Huffman,[55] samples are collected in organic-free glass containers. Two hundred milliliters are filtered on-site through a 0.45 μm porosity silver membrane filter to remove particulate and colloidal material; chilling to 4°C is the recommended method of sample preservation.

The method is applicable to water samples with DOC concentrations between 5 and 25 mg/ℓ and whose specific conductance is less than 2000 μmhos/cm at 25°C. Water samples with a DOC less than 5 mg/ℓ can be freeze-dried to the specific conductance limit; samples with DOC concentrations greater than 25 mg/ℓ should be diluted with organic carbon-free water to a DOC of 25 mg/ℓ prior to analysis. If the sample has a pH less than 6.5, the pH is adjusted to 7 by drop-wise addition of 1.0 *M* sodium hydroxide, since the hydrophobic bases become protonated in acid solution and these protonated bases will not adsorb on XAD resins.

The first step in the analysis is the measurement of the total DOC on a 10 mℓ aliquot of the water sample. The sample is adjusted to a pH of 7 and is then pumped at 2 mℓ/min. through an XAD-8 column containing 3 mℓ of XAD-8 resin and the hydrophobic bases and neutral solutes are removed from the sample. After the sample has been passed through the XAD-8 column, 0.1 *M* hydrochloric acid is pumped through the column to elute the hydrophobic bases. The procedure that is followed after the bases have been removed from the XAD-8 column is as follows: (1) the XAD-8 column is connected in series with a column containing 3 mℓ of AG-MP-50 BioRad® H^+ saturated cation-exchange resin (to adsorb hydrophilic bases) and 6 mℓ of AG-MP-1 BioRad® OH^- saturated anion-exchange resin (to adsorb hydrophilic acids) (2) the pH of the water sample is adjusted to a pH of 2 with concentrated hydrochloric acid, and (3) the sample is pumped through all three columns. The hydrophobic acids (now in the nonionized form) are adsorbed on the XAD-8 resin in this step and are subsequently eluted from the XAD-8 with 0.1 *M* sodium hydroxide and the DOC is determined. The hydrophilic acids and bases are eluted from the ion-exchange resins and their DOC is determined. The DOC of the hydrophilic neutrals is determined from the sample that has passed through all three columns. In this procedure the neutral hydrophobic fraction is not eluted from the XAD-8 column and is determined by difference. For the analyst who wishes to extend this method and characterize the individual components of the hydrophobic neutral fraction which contains the PAH, this fraction can be eluted from the XAD-8 with an organic solvent such as diethyl ether, methanol, or acetonitrile. The procedure can serve not only to separate the DOC into characteristic fractions but also to isolate the PAH, which are normally only a small fraction of the organic solutes found in water systems. Stuber and Leenheer[56] have used preparative scale DOC fractionation to study sorption of oil-shale retort-water organic solutes on processed oil shale.

The column distribution coefficient, k^1, defined by Thurman et al.[48] can also be used to estimate the concentration factor achieved by using Amberlite® XAD-8 columns and assuming elution with three pore volumes of eluant (which proved sufficient to elute the solute from all test solutions). The calculation is made with the following equation: concentration factor from original solution = $k^1/7.5$. Malcolm et al.[52] also use k^1 in the following emperical equation to calculate the complete retention of any solute

$$\frac{6(x)}{y} = k^1 \qquad (6)$$

where x = volume of sample in mℓ and y = bed volume of the column in mℓ and 6 is an empirically determined constant for XAD-8. Using this equation, the analyst can vary both the bed volume and the sample size depending on experimental conditions. For example, the following calculation is used to determine the volume of sample from which fluorene can be completely removed using a 20-mℓ bed volume of XAD-8. The solubility of fluorene is 1.69 mg/ℓ[24] = 1.0×10^{-5} mol/ℓ and log solubility is −5.0. Extrapolating the data in Figure 13, log k^1 = 2.3×10^4. From Equation 6, 6(x) = 7.7 $\times 10^4$ mℓ or 77 ℓ.

Since fluorene is one of the most water soluble PAH that will be encountered in water analysis, a bed volume of 20 mℓ that corresponds to ~ 6 g of XAD-8 is more than adequate to collect the PAH and other hydrophobic solutes from a 100-ℓ sample of water. Natural water may have a DOC value as high as 20 mg/ℓ and if the hydrophobic fraction is about one half of this value, the PAH must compete with other organic solutes for adsorption sites.

Another advantage of using the XAD resins is their ability to serve also as a desalting agent for isolating the hydrophobic fraction from inorganic salts which might interfere with certain instrumental methods.

As mentioned earlier, on-site extraction of pollutants from a water system has several advantages over collection of samples in the field which are then returned to the laboratory for analysis. The problems in transporting large sample volumes that are eliminated are the possibilities of chemical alteration and adsorption losses on the container surface. In addition, the continuous sampling of a water system should be expected to give a more representative sample than a grab sample. Stepan and Smith[38] have shown that in order to obtain quantitatively reproducible results when sampling with XAD resins it is necessary to control the pH of the water. This is obviously impractical with on-site sampling but can be measured with a pH meter. However, depending on the pH of the water system, some of the acidic or basic components may not be collected on the resin.

The XAD resin adsorption method has been widely tested by a number of workers both in the laboratory and in the field for the analysis of PAH and other organic pollutants in water systems. Shinohara et al.[50] have obtained recoveries of 95% using model systems containing 2,3 dimethylnaphthalene, fluorene, phenanthrene, pyrene, chrysene, and benzo(*a*)pyrene, B(a)P, at the ppt (pg/g) level using XAD-2 in an extraction system. By passing 200 ℓ of water through an XAD-2 column they were readily able to detect and analyze PAH and hydrocarbons at the ppt level. With a 200-ℓ water sample the detection limits are: acenaphthene 1.5 ppt, benzo(*a*)fluorene 0.1 ppt, dimethylnaphthalene 1.0 ppt, fluorene 0.02 ppt, phenanthrene + anthracene 0.05 ppt, benz(*a*)anthracene + chrysene + triphenylene 0.5 ppt, and benzo(a)pyrene 2.0 ppt.

The research group of Fritz, et al. at Iowa State University have studied trace organics in a number of water systems using both the laboratory and on-site extraction apparatus.[41,42] They use either sampling procedure, depending on whether instantaneous or an average concentration over a period of time is the point of interest. Composite sampling has the advantage that large volumes of water can be sampled and this is important when more of the contaminant is needed for identification purposes. The grab sample for laboratory extraction suffers the drawback of losses on the container walls during storage and transportation. In both of these methods, the eluant that is used to remove the adsorbed organics from the XAD must be reduced in volume during which step the volatile organic solutes are often lost. Those eluants that are not lost are generally masked by the solvent peak during gc analysis. The loss of volatile compounds can be reduced by use of the XAD mini-sampler. With the mini-sampler only 100 mℓ of water are sampled with 80 mg of resin (as opposed to 100 ℓ with 2 g of resin) and the sample can be eluted with 100 μℓ of solvent and no volume reduction is necessary. A second approach that can be used with the mini-sampler is to use the XAD column directly in a thermal desorption procedure. The EPA *Procedure for Analyzing Priority Pollutants* calls for head-space analysis to determine the low molecular weight volatile pollutants and we concur that this is the best procedure for recovering the highest percentage of the gc materials.

A number of specialized on-site sampling devices have been developed using XAD resins or Tenax-GC.® One such apparatus is that developed by Stepan et al.[60] The apparatus consists of a sampling module with two peristaltic pumps and a lead storage battery to power them. The operation of the pumps is controlled by an electronic timer so that the sampler can be operated either in the intermittent or continuous mode. A voltage stabilizer circuit allows the operator to vary the pumping speed between 0 to 60 mℓ/min. A 15 × 2 cm stainless steel cartridge contains the XAD-2 or XAD-7 resin. The resin is treated as described previously and 20 mℓ of the methanol slurry is loaded into the cartridge which is securely capped until it is installed in the sampler. In the field the cartridges are fitted into the sample holder. The sample holder serves to protect the cartridge from damage. A stainless steel mesh screen at the base of the cartridge prevents clogging of the resin by particulate matter. To avoid contamination from the connecting tubing and the pump, the cartridge is placed at the inlet of the sampling system and submerged. At the end of the sampling period, the cartridge is removed from the sampler, resealed, returned to the laboratory, and the solutes are eluted, concentrated, and analyzed. Tests carried out to compare the results of on-site extraction with grab samples showed the results to be identical to within 10%.

Since peristaltic pumps are equally effective with gases and liquids, the Stepan et al.[60] field sampler was also used for low-volume on-site extraction of organic compounds from air. For air sampling, Tenax-GC® was used as the adsorbent. The Tenax-GC® was contained in a 3 cm × 1 cm glass tube and kept in place by sintered glass discs. The Tenax-GC® was conditioned by passing helium through the Tenax-GC® while the cartridge was maintained at 200°C. After sampling, the adsorbed organics were eluted from the Tenax-GC® with 40 mℓ of hexane. The eluent was concentrated and analyzed by gas chromatography.

C. Open Pore Polyurethane

Polyurethane foams have been used by a number of investigators to recover PCBs, pesticides,[61-63] and phthalate esters from water.[64] These applications have prompted the investigation of the retentive properties of PAH on these foams.

Two groups have studied the removal of PAH from water systems with open pore polyurethane (OPP) foams. In their studies, Navratil et al.[65,66] prepared the OPP foam by polymerization directly in the sampling column, whereas Saxena et al.[34,67] have used commercially available foam plugs; diSPo® plugs (Scientific Products, Inc.), Identi® plugs (VWR Scientific) and plugs cut from UU34® foam sheets (Thomas C. Forrest Co., Inc.).

Navratil et al.[66] describe in detail the in situ preparation of the OPP columns. The OPP tubes are prepared by mixing 60 to 40 volume % solutions of an isocyanate and polyol into a glass column 5 or 10 cm long and 3.2 mm in diameter and allowing the polymerization to take place in the tube. The isocyanate used was Mondur MR® (Mobay Chemical Co.) which is a mixture of 4,4′-diphenylmethane-diisocyanate and some tri, tetra-, and pentaisocyanates. The polyol used was LA-475® (Union Carbide Co.) and it is primarily a pentahydroxy compound formed by the total oxypropylation of diethylenetriamine.

Navratil et al.[66] investigated a number of variables in the polymerization process.

They investigated monomer formulations with OH/NCO ratios of 1.0 and 2.2 and concluded, on the basis of breakthrough studies with pyrene, that the OPP prepared with OH/NCO = 2.2 had a much higher capacity than OPP with an OH/NCO = 1.0. Data were also presented to show that OPP polymerized at 0°C permitted a higher flow rate than the OPP polymerized at 25°C. Larger microspheres were formed when the reactants were polymerized at the lower temperature. We have used OPP tubes prepared according to the directions of Navratil et al. and have found, in addition, that the foam polymerized at 0°C maintain a better OPP-to-glass bond than those polymerized at 25°C.

The breakthrough capacity of the OPP columns was tested with pyrene test solutions and found to be superior to a number of resin-packed columns. The OPP formed from the OH/NCO = 2.2 formulation had a higher breakthrough capacity than BioRad® AG-MP-50 (BioRad Laboratories) and Amberlite® XAD-2.

Navratil et al.[66] eluted the PAH test compounds from the OPP columns with methanol and for a test solution containing B(a) P, fluoranthene, and pyrene; 90 to 98% recoveries were obtained.

Saxena et al.[67] studied the optimum parameters for the extraction of PAH from water using polyurethane foam plugs. In their studies, C^{14} labeled B(a)P was employed as a means of avoiding the necessity of determining the background levels of B(a)P in the water samples studied.

The extraction columns were prepared by wetting polyurethane foam plugs and inserting them in 25-mm Chromaflex® columns (Kontes Glass Co.). The prepared columns were washed successively with acetone, benzene, acetone, and distilled water. The columns were tested with a model sample of spiked water which was drawn through the column at a constant rate of 250± 10 mℓ/min at 25°C. The recovery data showed that the three types of foam plugs tested, diSPo®, Identi®, and plugs cut from UU34® foam sheets, performed equally well.

With the 25-mm columns the retention efficiency of B(a)P at ambient temperature remained unaltered for flow rates between 130 to 520 mℓ/min. A flow rate of 250 mℓ/min was selected, since at higher flow rates the plugs had tendency to slip to the bottom of the Chromoflex® column. Larger diameter columns were investigated but it was found that when the column diameter was increased from 20 tp 50 mm, the recovery of B(a)P decreased by 20%. Retention efficiency as a function of pH was studied using spiked tap water samples. At pH = 3.0 the retention efficiency was 59%, at pH = 6.7 it was 62 to 65% and increased to 76% at pH = 10.0. When the foam plugs are used to extract PAH from water samples in the laboratory the pH of the water can readily be adjusted for maximum efficiency, however for on-site extractions this is not convenient and the pH of the water must be measured so that the appropriate correction factor can be applied.

The effect of water temperature on extraction efficiency is most interesting and is shown in Figure 14. The retention efficiency of the foam plugs increased temperature up to 40°C, decreased, and then increased again with a further increase in temperature and reached a plateau at 60°C. On the basis of these measurements water samples were heated to 60°C prior to passage through the polyurethane packed column.

The possibility of increasing the retention efficiency of the foam plugs by coating them with a number of chromatographic phases was investigated.[67] The plugs were coated with DC-200, SE-30, and a nematic liquid crystal (*N,N′* bis(p-methoxybenzylidine) — α, α′ -bi-*p*-toluidine). Improvements in retention efficiency of between 4 to 9% were obtained. However, these improvements are nullified by the fact that when the sample is eluted from the foam plug the chromatographic coating is also eluted, thus complicating further concentration and analysis.

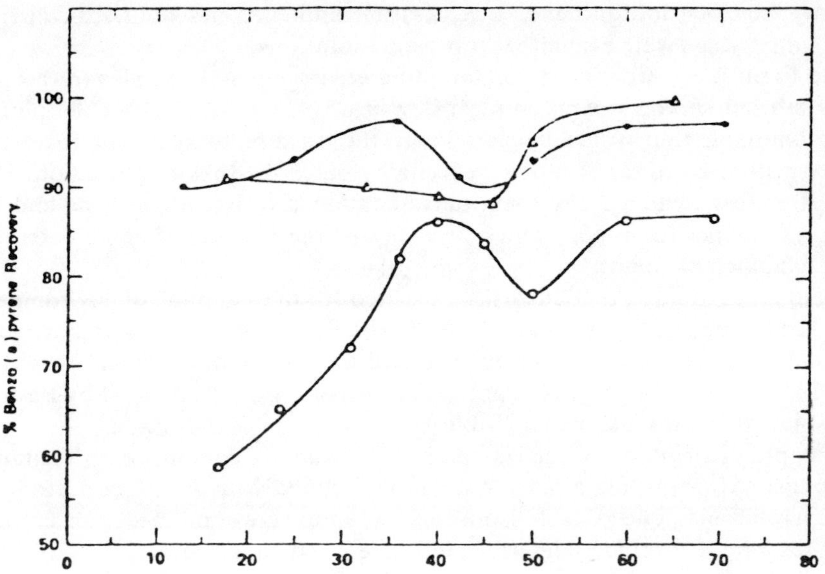

FIGURE 14. Effect of water temperature on extraction efficiency of PAH from polyurethane foam plugs. Water volume: 4 ℓ; B(a)P concentration, 0.1 ppb; flow rate, 250 ± 10 mℓ/min ○ : Tap water (unfiltered), ● : Filtered tap water, △ : distilled water. (Reprinted with permission from Basu, D. K. and Saxena, J., *Environ. Sci. Technol.*, 12, 791, 1978. Copyright by the American Chemical Society.)

Retention studies as a function of B(a) P concentration between 0.002 and 25 ppb were carried out using 4-ℓ samples and the recovery factory remained between 84 to 87% over this range of concentration.

The retention efficiency of the foam plugs remained constant with increasing volume of water sample when distilled water, free of extraneous solids or dissolved matter, was sampled. However, when successive increments in volume of tap water spiked with 0.05 ppb B(a)P were passed through a single plug, a steady decrease in retention efficiency was observed. The retention efficiencies for 4, 10, 20, and 40 ℓ of water were 87, 73, 67, and 49%, respectively. To overcome this drop in retention efficiency, the number of foam plugs was increased. When four foam plugs (two each in two different columns) were used, the retention efficiency for a 20-ℓ sample of spiked tap water was raised to 85%. The "four foam plug" sampler was tested with raw drinking water that contained 102 mg/ℓ of total suspended solids and 2.4 mg/ℓ of total dissolved solids and was spiked with 0.1 ppb B(a)P. When 4 ℓ were sampled with one foam plug the retention efficiency was 69% and the retention efficiency was increased to 81% when two foam plugs were used. On the basis of these experiments, Saxena et al.[67] concluded that two columns, each containing two foam plugs, should be adequate to concentrate the B(a)P from 20 ℓ of finished water and this configuration should also be used for sampling B(a)P from 10 ℓ of raw water.

When the foam plugs were stored at 4°C for 7 days,[67] no effect on the recovery of B(a)P was observed, but only 82% of the B(a)P was recovered when the plugs were stored at room temperature for the same period of time.

Basu and Saxena[34] extended these pilot studies with C^{14} labeled B(a)P to the six PAH recommended for monitoring by the World Health Organization.[5] Using the extraction column configuration described above, the sampling efficiencies for benzo(*a*)pyrene, fluoranthene, benzo(*j*)fluoranthene, benzo(*k*)fluoranthene, indeno(*1,2,3-cd*)pyrene, and benzo(*ghi*)perylene were evaluated with model solutions.

When the foam plugs were used for sampling environmental samples (rather than isotopically labeled spiked water samples) it was necessary to preclean the plugs to remove contaminants that would be eluted with the trace pollutants. The foam plugs were cleaned with cyclohexane and/or benzene by batch or Soxhlet extraction. However, even after this cleaning, the foam plugs contained 6 ng fluoranthene and 1 ng benzo(*ghi*)perylene per foam plug which necessitated the running of blank determinations with each batch of foam.

When foam plugs or any other adsorbent are used for the sampling of environmental samples, other organic pollutants in addition to the PAH are adsorbed and are subsequently eluted during the extraction step. Some of these compounds interfere with the PAH analysis and a cleanup step is required. The separation scheme used by Basu and Saxena to isolate the PAH is shown in Table 9.

A 60-ℓ sample of drinking water was passed through six columns each containing two foam plugs. After sampling, each column was eluted with 30 mℓ of acetone and 125 mℓ of cyclohexane. The extracts from each column were combined, concentrated, and the solutes were separated as described in Table 9 and analyzed.

The retention efficiency of the polyurethane foam plugs was also investigated with laboratory tap water spiked with PAH at the 25 ppb level.[34] The results of this experiment are shown in Table 10. The plugs maintained the same high efficiency when heavily polluted surface waters were sampled. The limit of detection of the six PAH when 60 ℓ of water were sampled using gc/FID is about 5 ng/ℓ.

For on-site sampling, Basu and Saxena[68] have employed a portable version of the polyurethane foam plug sampler to analyze groundwater, raw water, and treated waters. The field apparatus consisted of a pumping unit, a thermostated water circulator to maintain the water at 60°C, two columns each containing two foam plugs, and temperature and water flow monitoring devices. In this manner, the flow rate and water temperature conditions optimized in the laboratory could be maintained in the field. The columns were changed after every 20 ℓ when sampling treated waters and after every 10 ℓ when sampling raw water.

^{14}C-benzo(*a*)pyrene was added to the foam plugs as an internal standard before the extraction step. The recovery factor determined for this standard was used to correct the final results for losses in the elution and cleanup steps. The final result must also be corrected for the traces of PAH contributed by the foam plugs. Concentrations of individual PAH as low as 0.2 ng/ℓ were determined in drinking water, and concentrations as high as 600 ng/ℓ were observed in raw waters.

The use of labeled PAH is a direct and effective technique for testing the efficiency of an adsorbent for extracting PAH from water samples. The data presented by Basu and Saxena show that polyurethane foam plugs are effective in removing PAH from water, however the nuisance of exchanging columns during sampling and the large volumes of solvents required to elute the PAH from the foam plugs does not make this procedure attractive.

D. Carbon

The carbon-chloroform extraction method was developed by Braus et al.[69] to determine the efficiency of organic solute cleanup in drinking water plants. Carbon has been used for concentrating chlorinated insecticides and biphenyls. Rosen and Middleton[70] used activated carbon filters for determining the concentration of pesticides, and found that the results may be low by a factor of two. Hoak[71] studied the adsorption of phenol on activated carbon and found that while the adsorption efficiency was about 94%, the subsequent desorption was as low as 22%. Hoak also noted that chem-

Table 9
STEPS IN THE SEPARATION OF PAH FROM POLYURETHANE FOAM EXTRACT

Concentrated foam extract in cyclohexane (10 mℓ)

1. Wash with two 60-mℓ portions of 4:1 methanol/water.
   a. Retain cyclohexane layer.
   b. Discard methanol/water layer.
2. Wash with two 60-mℓ portions of distilled water.
   a. Retain cyclohexane layer.
   b. Discard aqueous layer.
3. Extract with three 20-mℓ portions of DMSO.
   a. Retain combined DMSO layer.
   b. Discard cyclohexane layers.
4. Add 120 mℓ of distilled water, then extract with two 40-mℓ portions of cyclohexane.
   a. Retain cyclohexane layers.
   b. Discard DMSO/water layer.
5. Dry the cyclohexane layer by passage through anhydrous sodium sulfate.
6. Concentrate to 5 mℓ.
7. Put on the Florisil® column and elute with 125 mℓ of benzene.
8. Concentrate the eluate to 0.5 mℓ.

Table 10
FOAM RETENTION EFFICIENCIES FOR PAH FROM TREATED WATER

Compound	Amt added to water (μg)	% Retention
Fluoranthene	100	100
Benzo(*j*)fluoranthene	100	88[a]
Benzo(*k*)fluoranthene	100	
Benzo(*a*)pyrene	100	81
Indeno(*1,2,3-cd*)pyrene	100	89
Benzo(*ghi*)perylene	100	91

Note: Water source: Laboratory tap water; water volume: 4ℓ; concentration of each PAH: 25 ppb; method of detection: gc-FID.

[a] Combined value given since the compounds could not be separated on the gc column.

ical changes in the adsorbed phenol could have affected the apparent efficiency of the carbon adsorption method. Eichelberger and Lichtenberg[72] have examined the carbon adsorption method for the recovery of organochlorine and organophosphorous pesticides from water samples. They recovered less than 10% of the fenthion and methyl parthion and were not able to recover any of the malathion, ethion, or methyl trithion.

Booth[73] found that closely reproducible results could be obtained for the comparison of total chloroform extractables using the carbon adsorption method. However, *The Standard Methods for Examination of Water and Wastewater,* 13th edition, notes that not all organic compounds are adsorbed on carbon and of those adsorbed not all are desorbed by chloroform.

Van Rossum and Webb[46] have recently carried out a careful investigation of the use of carbon alone and carbon in conjunction with XAD resins for the adsorption of organic pollutants from water. They included several PAH in their model systems. The problem of removing background impurities received careful attention in this study. The carbon used as the adsorbent was Filtrasorb® 300 (Pittsburgh Activated Carbon, Pittsburgh, Pa.).

The stock carbon was compared to carbon that had been cleaned by 400 chloroform extraction cycles in a Soxhlet extractor. The results of an in-column wash of one bed volume of acetone and three bed volumes of chloroform of these two carbon samples are shown in Figure 15 "first wash" and it can be seen that the results are almost identical. A second wash with three bed volumes of chloroform gave the acceptable blanks shown in Figure 15, "second wash". Van Rossum and Webb point out that the time consuming Soxhlet extraction step is unnecessary for the Filtrasorb® 300 when only small amounts of carbon and solvents are used in a water sampling system. The chromatograms shown in Figure 15 were obtained using a polar Carbowax® column. When the same precleaning rinses that were used to obtain the chromatograms shown in Figure 15 were analyzed on a nonpolar SE-30 column, the chromatogram shown in Figure 16 was obtained. This chromatogram indicates that no precleaning of the carbon is necessary if the solutes extracted from the water sample are to be determined on a SE-30 column. These experiments show that clean or unclean carbon must be defined in terms of the analytical system used. Van Rossum and Webb present data that shows that higher recoveries are obtained from carbon columns prepared with in-column prewashed material than with unwashed material. Acenaphthene was not recovered from a spiked tap water from either the washed or unwashed carbon when the carbon was extracted with a mixture of acetone and chloroform. In another experiment using carbon adsorbers, the carbon was extracted with benzene in a Soxhlet extractor and less than 20% of the fluoranthene and pyrene were recovered.

On the basis of these experiments Van Rossum and Webb[46] concluded tbat XAD-2 or a mixture of XAD-4/8 "is better in every respect than carbon." However, they used a carbon column as a backup for an XAD-4/8 column to sample 1000 ℓ of tap water to check if the carbon might retain some compounds that were not adsorbed by the XAD resin. The resin and carbon adsorption traps were eluted separately with acetone and chloroform and the eluates were analyzed by gc/ms. The analyses showed that only some halomethanes and long chain n-hydrocarbons broke through the XAD column. The carbon was further extracted with toluene in a Soxhlet extraction apparatus. This extraction confirmed that there was no breakthrough of PAH materials from the XAD resin.

In spite of these negative results with carbon, Grob[75] has successfully used carbon pellets for the adsorption of organic pollutants including PAH in conjunction with headspace extraction.

IV. LIQUID EXTRACTION

Mieure and Dietrich[76] have examined liquid extraction of organic pollutants from water systems in addition to headspace extraction and thermal desorption methods. They selected methylene chloride as the extraction solvent for the liquid extractions. They cited the following advantages of methylene chloride: (1) it is available in highly

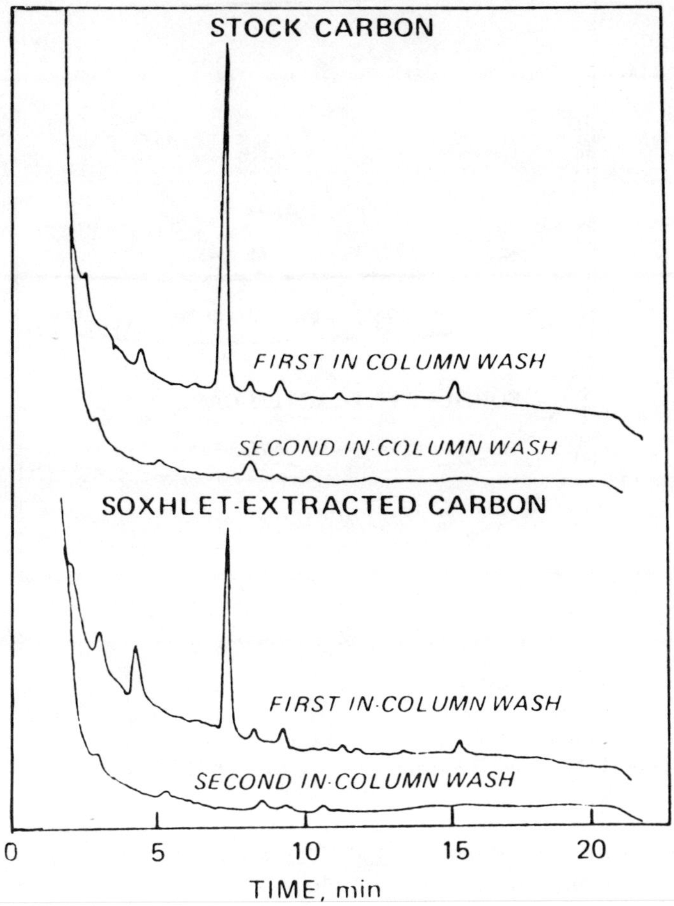

FIGURE 15. Comparison of background impurity levels of stock carbon and Soxhlet-extracted carbon by gas chromatography column: Carbowax® 20M - TPA. Temperature: 3 min at 80°C; 8°C/min to 170°C.

purified form, (2) it is relatively chemically inert, (3) it will dissolve many compounds at low concentration, and (4) since it is denser than water, it is readily separated from the water phase in a separatory funnel.

In the Mieure and Dietrich procedure a 500 mℓ sample of water is placed in a separatory funnel and 150 g of sodium chloride is added to form an almost saturated solution. The high concentration of the electrolyte in the water improves the extraction efficiency of water soluble compounds by a salting-out effect. Fifteen milliliters of methylene chloride are added, the funnel is shaken, and when the phases have separated the methylene chloride phase is drained through a cotton plug into a receiving vessel. Allowing for the solubility of methylene chloride in water about 10.0 ± 0.3 mℓ are recovered and if the extraction is complete a 50-fold concentration of the organic pollutants in the water has been achieved. The enrichment can be further enhanced by evaporating the methylene chloride, but this step involves the risk of losing volatile compounds.

Some organic compounds can be almost completely extracted from water with methylene chloride in one extraction step, while others are quite soluble in water and are

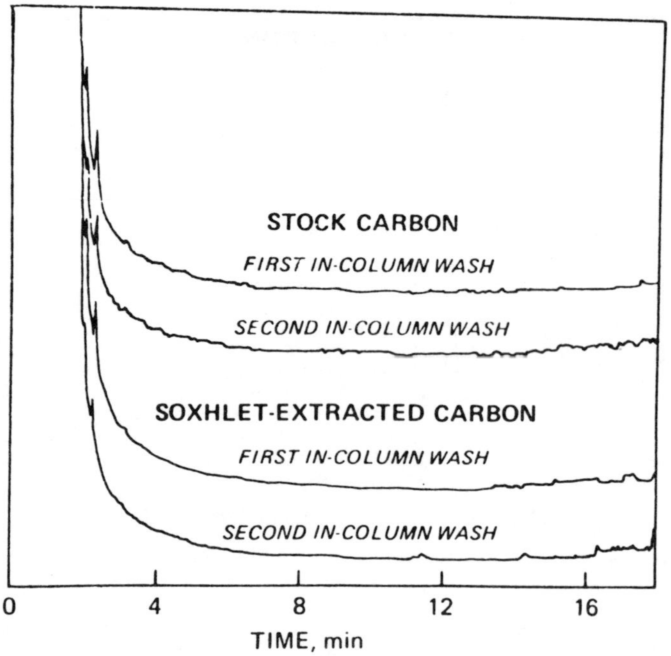

FIGURE 16. Comparison of background impurities. Levels on stock carbon and Soxhlet-extracted carbon by gas chromatography with SE-30 column. Temperature 100 to 225°C at 7°C/min.

only incompletely extracted. Pancirov et al.[77] have measured the efficiency of the direct extraction of PAH from a wastewater sample spiked with C^{14}(B(a)A and B(a)P) using methylene chloride as extraction solvent. The first extraction recovered 88% of the PAH and by the third extraction 98% of the PAH had been recovered.

When quantitative analytical results are needed, a correction must be applied to account for incomplete extractions. The most direct way to arrive at this correction is to run two sequential extractions on the same water sample. The second addition of methylene chloride can be reduced to 10.0 mℓ since the water is already saturated with methylene chloride. If the chromatographic peak area for a component in the first and second extractions are represented by A_1 and A_2, and the extraction efficiency is assumed to be the same for the first and second extraction, the total amount of A in the water sample can be calculated from the equation,

$$\text{Total A in water sample} = \frac{A_1^{\,2}}{A_1 - A_2} \qquad (7)$$

When samples of spiked tap water were extracted twice and analyzed, and Equation (7) was applied, satisfactory results were obtained at the ppm level even when the percentage recovered in the first extraction was less than 30%. A second approach to determine the total amount of A in the water sample is to extract the sample several times, combine the extracts, and analyze them. A third approach is to prepare and analyze standard samples. The latter method can be very time consuming for multicomponent solutions.

Grob et al.[78] have also examined the potential of liquid extraction for the recovery of organic pollutants from water at concentrations of 10 ppt (10 pg/g). As an example, a 1-ℓ sample is extracted with 100 mℓ of an organic solvent. If the extraction efficiency is assumed to be 100%, and if the extract is then reduced to 2 μℓ for a single injection onto a capillary column, the solvent is concentrated by a factor of 50,000. However, impurities in the solvent are also concentrated by this factor so that the resulting chromatogram contains both the original impurities in the water and in the solvent. Grob et al.[78] conclude that the combined effect of the losses of the pollutants and the enrichment of the solvent impurities make this procedure impractical. To avoid the problems created by large volume reductions, they developed a unique liquid extraction method. In this method, 1 ℓ of water is extracted by 220 μℓ of solvent, so that very little if any further concentration is required before the sample is injected into the gc. To achieve a reasonable extraction with so small a volume of solvent, a solvent of very low aqueous solubility is necessary, so that the fraction of solvent remaining in the aqueous phase is minimized. Furthermore, any solvent dissolved in water solubilizes trace levels of organic material. This effect becomes the important factor in the distribution of organic substances between the phases at extreme trace level concentrations. As a result, distributions that cannot be predicted from simple distribution coefficient considerations may result. The solvents that best meet these requirements are the higher alkanes, but pentane was selected as the best compromise between low solubility, reasonable equilibration time, and a low gas chromatographic retention time, which allows the more volatile water pollutants to be detected without solvent masking.

Grob et al.[78] used a series of even-numbered 1-chloroalkanes as internal standards and these were added to the water samples containing the unknown pollutants. The advantages of using this homologous series as internal standards are twofold; (1) the compounds of the series are generally absent in polluted waters and are readily identified on the gas chromatogram, and (2) recovery factors as a function of the varying volatility of extracted substances can be interpreted. To carry out the extraction, 0.6 mℓ of pentane is added to 900 mℓ of the water sample at 12°C in a 1-ℓ volumetric flask and the flask is shaken vigorously by hand for 2 min. More water is added to raise the liquid level into the narrow neck. One milliliter, containing the pentane and some water, is removed and transferred to a 15 × 4 mm glass tube. The organic phase, which is about 200 μℓ, is transferred from this tube into a conical test tube. In the case of heavily polluted waters, the extract can be analyze directly, or in the case of very pure waters the volume of the extract is reduced to 3 μℓ by evaporation. The losses of volatile solutes are gauged by the internal standards.

The results of this liquid extraction procedure are compared with the results obtained from headspace analysis using a water sample spiked with 2.5 ppb of diesel oil. As might be expected, liquid extraction is only one third as efficient as headspace analysis for low to medium molecular weight substances; at C_{20} (eicosane), both methods are equivalent, and for high molecular weight substances liquid extraction is more efficient.

Murray[79] has developed an extraction flask for liquid extractions with small volumes of solvent using Grob's method.[78] In place of the 1-ℓ volumetric flask, Murray has used a two-neck 1-ℓ flask with a capillary tube sealed to a shortened center neck. Nine hundred eighty milliliters of water and 200 μℓ of hexane are shaken manually for 2 min. The extraction flask is then tilted, so that by the addition of more water through the sidearm the solvent layer is forced into the capillary and can be readily recovered.

Acheson et al.[35] have compared conventional liquid extraction using a separatory funnel, continuous solvent extraction, and the Ultra-Turrax® (an ultrasonic mixer/homogenizer) method. They found that extraction in a separatory funnel gave low results and rejected this method.

For water samples free of suspended solids the Ultra-Turraxrm method and continuous solvent extraction method gave comparable recoveries (∿84%) for pyrene at a concentration of 0.3 μg/ℓ. For benzo(*ghi*)perylene at a concentration of 0.1 μg/ℓ, the Ultra-Turrax® method gave a recovery of 82% but only 55% was recovered by continuous solvent extraction. When Fuller's earth was added to the water as a suspended solid, the extraction efficiency by both methods dropped but the recovery by the Ultra-Turrax® method was significantly better than by solvent extraction. On the basis of these experiments, Acheson et al. selected the Ultra-Turrax® method for all their tests.

Acheson et al.[35] investigated several factors that influence the extraction and analysis of PAH in environmental water samples spiked with pyrene and benzo(ghi)perylene. Their results showed that extraction efficiency varies seriously with both the concentration of the PAH in the water sample and with the amount of suspended solids in the water.

The effect of prolonged mixing of water samples, both with and without suspended solids, spiked with PAH and stirred in the dark for 6 hr was investigated. The 6-hr mixing lowered the extraction efficiency equally for samples with and without suspended solids. Acheson et al.[35] postulate, and we concur, that these results can be explained by adsorption of PAH onto the walls of the mixing vessel. They conclude, as we have noted previously, that it is important to sample water directly into the extraction vessel and carry out the extraction as quickly as possible.

Harrison et al.[80] have studied the effect of water chlorination on the levels of some PAH in water. For this study they used 5-ℓ samples of water spiked with eight PAH and employed the methylene chloride/Ultra-Turrax® extraction method of Acheson et al.[35] They added 300 mℓ of distilled methylene chloride and emulsified the samples for 5 min in the Ultra-Turrax® apparatus. After standing in the dark overnight, the methylene chloride layer was separated, concentrated, and analyzed by gc. Using octacosane as an internal standard, recovery efficiencies for the eight PAH ranged from 64 to 68%. Harrison et al.[80] studied the effect of a time delay on the extraction efficiency of the PAH from water. A standing time of 45 min did not affect the recovery efficiency of benzo(*a*)pyrene and perylene, decreased the recovery of pyrene, fluoranthene, benz(*a*)anthracene and benzo(*k*)fluoranthene by 10%, and decreased the recovery of indeno(*1,2,3-cd*)pyrene and benzo(*ghi*)perylene by 20%. The concentration of free chlorine, the temperature, and pH of the water also affected the recovery of PAH.

Particulate matter was found to significantly affect the PAH level of the water. When larger particles settled out, only a small reduction of the PAH level in the water was observed, but a 50% reduction in PAH occurred when the smaller particles were removed by filtration. These observations of Harrison et al.[80] are in accordance with the estimate of Reichert et al.[81] that 60% of the PAH in river waters exist adsorbed on particulate matter. Borneff and Fischer[82] have noted that the backwash mud from sand filters contains high levels of PAH. Paralleling these observations, Harrison et al.[9] have previously reported that filtration of water through beds of activated carbon removed up to 99% of the PAH.

Continuous liquid extraction overcomes two limited factors in solvent extraction: (1) the saturation capacity of the solvent is eliminated by continuously providing fresh solvent, and (2) the volume of liquid available for extraction is not limited as in batch extraction. Continuous liquid extractors designed to use either heavier-than-water or lighter-than-water solvents have been described by Goldberg et al.[83] The extractors consist of a two-cycle system: a water cycle and a solvent cycle. The extractor that uses lighter-than-water solvent is shown in Figure 17. Water enters at A and exists at B passing through the chamber C which is half-filled with solvent. The solvent cycles exclusively in the extractor. The bulb C contains the solvent. When the bulb is gently heated the solvent vaporizes, rises through tbe apparatus, is condensed at H, and drops into the funnel J. The hydraulic head in the funnel stem forces the solvent through

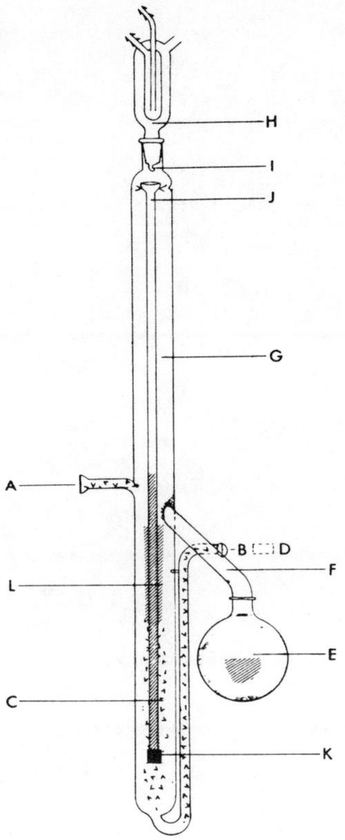

FIGURE 17. Extractor design for lighter-than-water solvents. The designated parts are described in the text.

the fritted disc K. The frit homogenizes the solvent which results in fine bead-like particles which form an emulsion as they rise through the water in chamber C, extracting organic solutes. The emulsion separates at L and the solvent-solute mixture spills over into the flask E.

The heavier-than-water solvent extractor is shown in Figure 18. Water enters at M and exits at N. While in the lower section of the extractor P, the water flows through the solvent. The solvent cycle is as follows: the solvent is vaporized in Q, condenses at S, and drops into the funnel U. In the funnel a hydraulic head is created and this forces the solvent through the frit V and it drops into the upper part of the extractor W. Extraction of the organic solutes takes place at the interface between the emulsified solvent and the water. A stirring bar X stirs the solvent-water mix. The solvent separates in the lower section of the extractor and flows through the tube Y into the bulb Q. The extracted organic solutes are concentrated in flask Q.

The extractors were tested by setting up three solvent heavier-than-water extractors in one series and four solvent lighter-than-water extractors in a second series. Each extractor was charged with 500 mℓ of solvent, and a continuous feed of test water was passed through the extractors. Data were presented for a number of solute-solvent combinations.

Since a large volume of solvent is involved in this system, the extraction is followed by a volume reduction step with a Kuderna-Danish apparatus. This step involves the

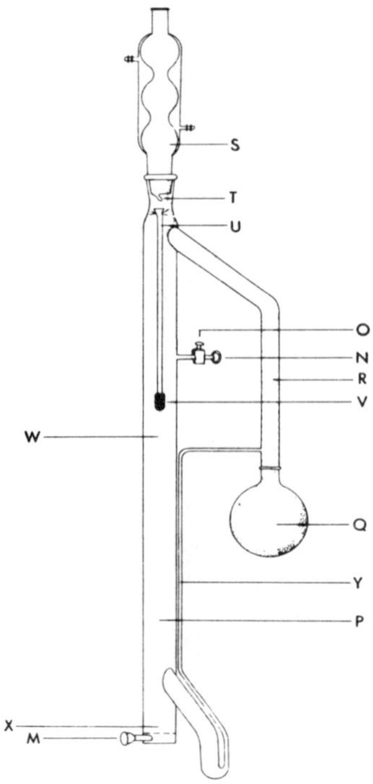

FIGURE 18. Extractor design for solvents heavier-than-water. The designated parts are described in the text.

risk of concentrating impurities in the solvents as well as losing some of the lighter solutes. However, concentration factors up to 10^5 can be achieved with the system. The dipole moment difference between the solvent and solute was shown to be an index of extraction efficiency for liquid-liquid extraction. Kahn and Wayman[84] have also described a continuous extractor using lighter-than-water organic solvents.

Another continuous liquid-liquid extraction apparatus has been described by Ahnoff and Josefsson.[85] The apparatus is in principle a mixer-settler in which the mixing and settling chambers are not completely separated but combined in one cylinder. The efficiency of this extractor was tested by analyzing standard solutions in three extractors connected in series. In these tests cyclohexane was used as the solvent and standard solutions of pesticides prepared in triple-distilled water were pumped through the three extractors at pump rates of 1.1, 2, and 5 ℓ/hr. At a pumping rate of 1.1 or 2 ℓ/hr the extraction efficiency averaged between 89 to 96% and dropped to 83 to 86% when the pump rate was increased to 5 ℓ/hr.

The apparatus was field tested with water volumes of 175 to 305 ℓ, using series and parallel arrangements at flow rates between 3 and 5 ℓ/hr. The extracted solvent was cleaned and concentrated and analyzed with an ECD gas chromatograph; PCB concentrations in the range 0.1 to 1.0 ng/ℓ were reported.

More recently Ahnoff and Josefsson[86] have published the description of an integrated sampler which combines two extractors of the type described above placed in series. Data is given for the extraction and analysis of PCB in estuarine water at a dredging sludge disposal area.

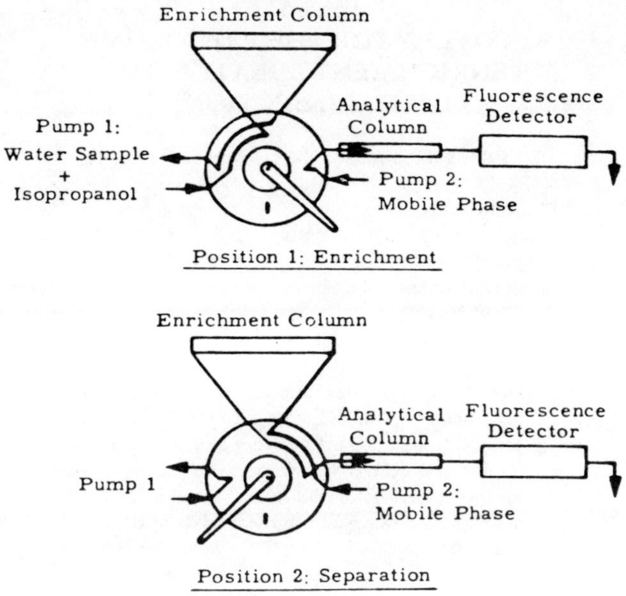

FIGURE 19. Configuration of components for the two positions of the injection valve for the separation and determination of PAH by hplc. (From Ogan, K., Katz, E., and Slavin, W., *J. Chromatogr. Sci.*, 16, 517, 1978. With permission of Preston Publications, Inc.)

V. COUPLED-COLUMN LIQUID CHROMATOGRAPHY

The principle of coupled-column/high performance liquid chromatography has been used by a number of investigators to concentrate and analyze PAH from water systems. In this method the PAH are enriched on an adsorption column, directly eluted from this adsorption column onto an analytical column, and separated by reversed phase high performance liquid chromatography. The advantages of this method are: (1) the volume of the sample is minimized, since all the PAH in the water sample that are extracted are transferred to the analytical column and determined, and (2) sample handling for extraction and concentration is eliminated.

A system of this type has been described by Eisenbeiss et al.[87] A block diagram of the enrichment and analytical system is shown in Figure 19. The water sample is initially pumped through the enrichment column. At the end of the sampling period, the four-port valve is rotated and a second pump forces the mobile phase through the enrichment column. The PAH are eluted and transferred to the analytical column where they are separated.

The packing of the enrichment precolumn must adsorb the PAH from the water sample and readily desorb the PAH when the mobile phase used in the analytical column is passed through it. Eisenbeiss et al.[87] found that no commercially available packing would fulfill both these requirements. They developed a special packing for the precolumn which is identified only as a "Special Merck packing". To achieve enrichment on this collection column, it was necessary that the water sample contain 17.5 to 20% isopropanol. This configuration was tested with a 500-mℓ test solution containing 17.5% isopropanol, 100 ng fluoranthene, 20 ng B(b)F, 20 ng B(k)F, 20 ng B(a)P, 20 ng benzo(*ghi*)perylene, and 20 ng IP, and the recovery factors are listed in Table 11.

Ogan et al.[88] have reported a similar analytical approach to the determination of

Table 11
RECOVERY FOR THE PRECOLUMN ENRICHMENT RELATIVE TO DIRECT INJECTION[a]

PAH compounds	Recovery rates %
Fluoranthene	100
Benzo(*b*)fluoranthene	85
Benzo(*k*)fluoranthene	75
Benzo(*a*)pyrene	76
Benzo(*ghi*)perylene	86
Indeno(*1,2,3-cd*)pyrene	85

[a] Determined as the ratio of chromatographic peak heights obtained from a 10 mℓ PAH standard solution in 500 mℓ of water and concentrated on the precolumn before analysis to the peak heights obtained by the direct injection of a 10 mℓ standard sample onto the analytical column.

From Eisenbeiss, F., Hein, H., Joester, R., and Naundorf, G., *Chromatogr. Newsl.*, 6, 8, 1978. With permission.

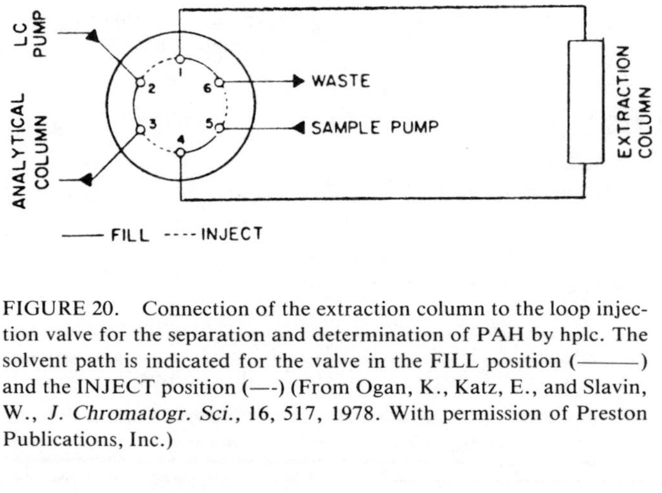

FIGURE 20. Connection of the extraction column to the loop injection valve for the separation and determination of PAH by hplc. The solvent path is indicated for the valve in the FILL position (———) and the INJECT position (—-) (From Ogan, K., Katz, E., and Slavin, W., *J. Chromatogr. Sci.*, 16, 517, 1978. With permission of Preston Publications, Inc.)

PAH in water. In their procedure they use a precolumn 10.0 × 0.26 cm packed with 40 μm C_{18}-bonded phase pellicular material, ODS-Sil-X-11, for the extraction of the PAH. About 800 mℓ of a model solution containing 20% methanol, and the five PAH, 7H-benz(*de*)anthracene-7-one (BA-one), fluoranthene, perylene, benzo(*a*)pyrene, and benzo(*ghi*)perylene were used to measure recovery factors and the sensitivity of the method.

The extraction and analytical apparatus used by Ogan et al.[88] is shown in Figure 20. In operation, the injection valve is turned to the INJECT position and 400 mℓ of the sample are pumped through the sample pump to equilibrate the pump and lines with the sample. The sample pump is turned off, the valve is turned to the FILL position, and 100 mℓ of sample are pumped through the extraction column and out to waste. After 100 mℓ have been pumped through the extraction column, the sample pump is stopped, the valve is turned to the INJECT position again, the LC pump is started,

Table 12
RECOVERY EFFICIENCY OF PAH BY THE INTEGRATED HPLC TECHNIQUE

Compound	Recovery %
7H-Benz(*de*)anthracene-7-one	97
Fluoranthene	101
Perylene	98
Benzo(*a*)pyrene	94
Benzo(*ghi*)perylene	96

From Ogan, K., Katz, E., and Slavin, W., *J. Chromatogr. Sci.,* 16, 517, 1978. With permission of Preston Publications, Inc.

and the extraction column is backflushed with the mobile phase and the PAH are transferred onto the analytical column. Since only 500 mℓ of the water sample have been used up to this point, a second 100 mℓ sample is pumped through the extraction column while the first chromatogram is eluting from the analytical column. With this procedure a duplicate can be run as soon as the first chromatogram has eluted. The percentage recovery of the individual PAH, as measured by the ratio of chromatographic peak areas of diluted samples to the areas obtained when the samples to the areas obtained when the sample was directly injected into the chromatograph, is uniformly high and significantly better than the recoveries obtained by most other methods (Table 12). The method of Ogan et al. when used with a 400 mℓ sample allows detection of PAH in water samples at levels as low as 0.2 ng/ℓ and is linear up to the μg/ℓ range of heavily polluted waters.

In all the procedures for sampling, concentration, and analysis reported and used, the adsorption of PAH on glass and metal parts can seriously affect the recovery efficiency. Ogan et al.[88] found that the flushing of all the components with the sample to saturated the adsorption sites and silanization of the glass significantly improved recoveries.

Other investigators[89,90] have also reported on the use of an hplc sampling and analytical procedure for the analysis of PAH in water. These other procedures do not include the addition of alcohol to the water sample before it is pumped through the precolumn. Eisenbeiss et al.[87] reported that without the addition of isopropanol, there was poor wettability of the precolumn packing and no enrichment of PAH. Ogan et al.[88] reported that the addition of methanol greatly reduces adsorption on glass and metal surfaces and recovery is improved. They state that "methanol undoubtedly improves the solubility of the PAH in the aqueous solution. Also it may enhance the interaction of the PAH with the extraction column packing ("wetting" the surface), or it may directly modify the packing surface."

May et al.[89,90] used coupled-column chromatography to analyze the water for PAH remaining after headspace analysis. They used a 6.5 × 0.6 cm o.d. stainless steel precolumn packed with Bondapak® C_{18}, a 37 to 50 μm pellicular support with a bonded C_{18} stationary phase. The water sample was pumped through this column at the rate of 10 mℓ/min. After the sample was passed through the precolumn, the precolumn was attached to a 30 × 0.6 cm analytical column that was packed with μ-Bondapak® C_{18}, a 10 μm microparticulate support with a bonded C_{18} stationary phase, and the sample was eluted with a mobile phase that was programmed from 30:70 methanol/water to 100% methanol. The recovery factor for five PAH by this procedure are shown in Table 13.

Table 13
RECOVERY FROM WATER (2 ℓ) OF INTERNAL STANDARD COMPOUNDS USING COUPLED-COLUMN LIQUID CHROMATOGRAPHY

Compound	Concentration (μg/kg)	Recovery %
Naphthalene	28	19±2
Phenanthrene	1.5	92±12
Pyrene	2	78±17
Benzo(*a*)pyrene	1.5	58±12
Dibenzanthracene	3	14±8

It can be seen that the recoveries are not as high as those reported by Eisenbeiss et al.[87] and Ogan et al.[88] where an alcohol was added to the water sample before it was pumped through the preconcentration column.

Oyler et al.[91] have also used coupled-column hplc in a study to determine the aqueous chlorination reaction products of PAH, and they also did not add alcohol to the water. Their reported recovery factors range from 36% to 115%. In this study, however, pyrene (MW206) was the highest molecular weight PAH used in the model solutions.

Euston and Baker[92] have reported several procedures in which the automatic features of the Hewlett-Packard® model 1084B liquid chromatograph can be used to include a concentration step as part of the analytical cycle. One option is to replace the water, which is normally used as the "weaker" component of the mobile phase in reversed phase chromatography, by the water sample and to concentrate the impurities on the colum in this manner. The column is first cleaned by pumping 100% acetonitrile through it for 5 min. At the end of this period, the mobile phase is changed to the environmental or spiked water sample, which is pumped through the column for 50 min at 2 mℓ/min. In this manner, the nonpolar impurities from the 100 mℓ water sample are concentrated at the inlet of the column. At the end of the sampling period, the composition of the mobile phase is adjusted to elute the PAH. The sharp peaks in the resulting chromatogram indicate that there was only negligible migration of PAH on the column during the 50 min loading period.

A second option described by Euston et al. is an automated version of the procedure used by Eisenbeiss et al.[87] and Ogan et al.[88] An automatically controlled six-port valve first switched the water sample to a 5 cm × 2.1 mm i.d. precolumn packed with Perisorb® RP (a superficially porous reversed-phase material), and out to waste. After the desired volume of water had been sampled, the six-port valve was automatically switched so that the mobile phase backflushed the solute that are adsorbed on the precolumn onto the analytical column. No data for recovery factors were given. The apparatus has a sensitivity for PAH in the range of 70 pg to 1 ng depending on the molar absorptivity of the individual PAH.

Huber and Becker[93] also used a high pressure liquid chromatograph to enrich water samples by displacement chromatography. The water sample was pumped through the enrichment column and the trace components adsorbed on the stationary phase. The trace components were subsequently shifted to the end of the column by the displacer fluid and measured either in-line or collected for off-line measurement. The enrichment of dibenz(*a,h*)anthracene is described. LiChrosorb® RP-18 (5 μm) was used as the column packing and dioxane was used as the displacer. A 250-mℓ sample of an aqueous saturated solution of dibenz(*a,h*)anthracene was extracted and concentrated to 0.2 mℓ, corresponding to an enrichment factor of 1250. The applicability to field sampling was investigated by loading samples on the columns and storing them for 3 days before displacement. Only slight tailing of the enriched sample was observed.

VI. HEADSPACE ANALYSIS

In headspace analysis, nitrogen or another inert gas is bubbled through or over the water sample and the inert gas, water vapor, and organic solutes are passed through a resin or cryogenic trap. When a hydrophobic resin trap such as Amberlite® XAD-2 or Tenax-GC® is used, the organic compounds are adsorbed on the resin and the inert gas and water vapor pass on through the trap. The advantages of headspace analysis are as follows:[76]

1. Many handling problems such as the filtering of particulates are avoided.
2. The technique can be readily automated for continuous monitoring.
3. No solvent extraction is involved so that low molecular weight, volatile compounds are not masked by the gc solvent peak and left undetected.

On the other hand, there is the disadvantage that nonvolatile, very insoluble pollutants, such as the heavier PAH, or pollutants that are adsorbed on particulates will not be completely removed from the water sample, giving rise to low results, whereas a pollutant concentrated in a surface film will give high results.

Zlatkis[94,95] used a small tube filled with Tenax-GC® to analyze trace volatile metabolites in serum and plasma. The organic substances collected on the Tenax-GC® were thermally desorbed, collected in a cold trap, and determined by gas chromatography.

Mieure and Dietrich[76] have used headspace analysis to determine the organic pollutants present in water. The method was used for substances that had boiling points less than 250°C and concentrations in the ppm range. The apparatus consisted of a 125-mℓ filter flask in which a 10-mℓ water sample was placed. Tenax-GC® or another solid adsorbent was used in the collection column. Laboratory air purified by passage through a charcoal trap was used to sweep the organics onto the collection column. Good reproducibility was achieved when the temperature of the flask was held constant at 25 ± 1°C. Because detection limits in this method are a function of the volatility of the solute in a 10 g water sample, they range between 1 ppm for relatively nonvolatile solutes to 1 ppb for volatile solutes. Sensitivity of the method can be increased or the analysis time shortened by heating the sample. Mieure and Dietrich point out that a 25°C increase in water temperature results in a 20-fold increase in the concentration of the volatile organics over the water. The analysis time can also be reduced if the inert gas is bubbled through the water rather than passed over it.

May et al.[89,90,96-98] have developed a procedure for the determination of hydrocarbons in marine sediments and seawater. The procedure uses headspace analysis for the light hydrocarbons and low molecular weight PAH, and coupled-column liquid chromatography for the nonvolatile compounds, which include the high molecular weight PAH. Water samples are collected in the field, an internal standard consisting of a solution of aromatic hydrocarbons is added, and the sample is stored in a freezer at −10°C. Prior to analysis the samples are thawed overnight in a laminar flow hood in a room maintained at 4°C. After thawing, approximately 750 mℓ of water or 100 g of sediment are transferred to the headspace sampling flask. In the case of the sediment samples, approximately 600 mℓ of hydrocarbon-free water (distilled from potassium permanganate and passed through an XAD-2 column) and the internal standard are added to the 100 g of sediment in the headspace analysis apparatus. The headspace analysis is carried out with scrubbed nitrogen in the apparatus shown in Figure 21.

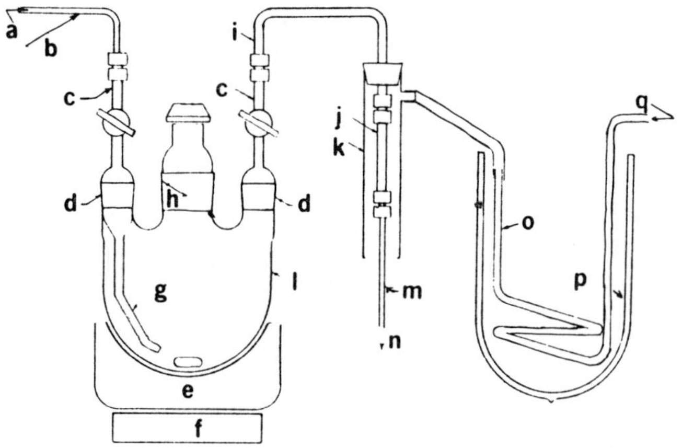

FIGURE 21. Headspace sampling vessel. (a) N_2 gas inlet, (b) copper tubing, 1/8 in., (c) glass tubing, 1/4 in., (d) tapered joint 24/40, (e) heating mantle, (f) magnetic stirrer, (g) dispersion tube with coarse frit, (h) tapered joint 45/50, (i) glass transfer line (1/4 in.), (j) TENAX® trap, (k) glass cooling jacket maintained at 15°C with air, (l) 2-ℓ round bottom flask, (m) Teflon® tubing, 1/8 in., (n) to rotometer, (o) copper coil, (p) ice bath, and (q) compressed air inlet.

Sampling is carried out for 2 hr at room temperature, then the flask is heated to 70°C and sampling is continued for a further 2 hr. The Tenax-GC® trap is cooled with a stream of air passed through a copper coil immersed in an ice bath. At the lower temperatures there is less likelihood of losses of adsorbed material from the Tenax-GC.® At the end of the sampling period, the Tenax-GC® collection column is disconnected from the apparatus, cooled, and dried for 2 hr by a stream of nitrogen gas at a flow rate of 150 mℓ/min. No data were given for losses of adsorbed material during this step.

The dried Tenax-GC® traps are capped and stored at 4°C. The materials adsorbed on the trap are determined by installing the trap as a precolumn between the injection port and coiled glass SE-30 coated SCOT column shown in Figure 22, and the sample is transferred to the gc column by flash heating.

After the volatiles have been extracted from the water or sediment sample by headspace analysis, the water remaining in the flask is pumped through a liquid chromatography precolumn which is subsequently attached to a liquid chromatograph capable of providing gradient elution.

The efficiency of the headspace analysis method can be assessed from the internal standard data listed in Tables 14 and 15 which show the recovery factors obtained from the headspace/gas chromatography analysis. The improvement in recovery factors when the sampling time is increased is shown in Table 15.

The Stepan and Smith method[38] is similar to that of May et al. The water sample is placed in the three-neck flask of the headspace apparatus, the flask is placed in a water bath thermostated at 95 ± 0.3°C, and nitrogen is bubbled through the sample at 10 mℓ/min. The trace organics are collected on a resin adsorption column that contains 2 g of XAD-2. The extraction time is 1 hr. Stepan and Smith were able to increase the extraction efficiency from 68 to 78% by placing the 2 g of XAD-2 in a 75 cm × 0.4 cm i.d. coil rather than a 8 cm × 1.2 cm cartridge. They were able to increase the recovery of ethylbenzene to 89% when the temperature of the coil containing the XAD-2 was

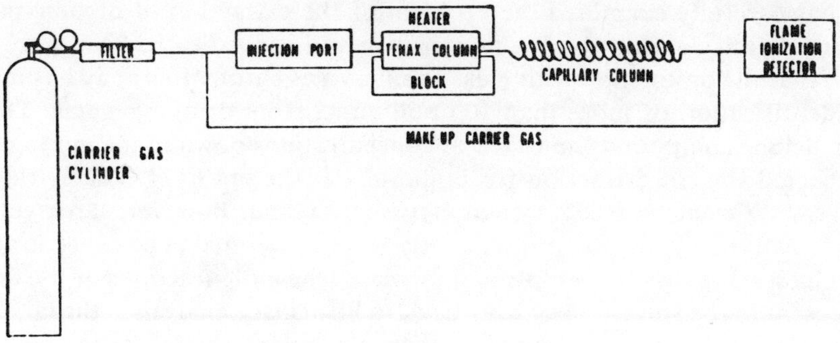

FIGURE 22. Schematic of gas chromatographic flow system showing location of TENAX-GC® pre-column.

Table 14
RECOVERY FROM WATER (600 mℓ) OF INTERNAL STANDARD COMPOUNDS USING HEADSPACE SAMPLING AND GAS CHROMATOGRAPHY

Compound	Concentration (μg/kg)	Recovery %
Mesitylene	2.7	8 ± 5
Naphthalene	3.5	38 ± 7
n-Propylnaphthalene	3.3	51 ± 10
Phenanthrene	3.3	19 ± 12

Table 15
PERCENT RECOVERY OF AROMATIC AND ALIPHATIC HYDROCARBONS FROM WATER BY HEADSPACE SAMPLING AND GAS CHROMATOGRAPHY

Compound	4-hr sampling	18-hr sampling
Mesitylene	8 ± 2[a] (6)[b]	6 ± 1 (3)
Naphthalene	29 ± 4 (6)	52 ± 9 (3)
2,3,6-Trimethylnaphthalene	—	95 ± 5 (3)
Phenanthrene	12 ± 4 (6)	92 ± 4 (3)
Pyrene	—	82 ± 4 (4)
5-Methyltetradecane	62 ± 4 (6)	84 ± 3 (3)
2-Methyloctadecane	57 ± 8 (6)	94 ± 3 (3)

[a] Data reported as the standard deviation of a set of replicate values from the mean of the replicate values.
[b] Denotes number of samples.

reduced from 20°C to 3°C. A similar improvement in recovery by cooling the polymeric adsorbent was also noted by Chesler, May, et al.[89,90]

The resin was extracted in the standard manner with one 20-mℓ portion and one 10-mℓ portion of diethyl ether. The eluates were combined, dried over magnesium sulfate, and analyzed by gc.

Grob[75] has carefully examined the problem of the extraction of organic pollutants from water. He has developed a highly sensitive and sophisticated headspace analysis procedure, that in conjunction with glass capillary gas chromatography has made possible the identification of more than 100 pollutants from drinking water. These contaminants include compounds up to C_{24} at concentrations down to the ppt (pg/g) level.

Grob rejected solvent extraction for pollutants at the ppt level. The extraction can be made very efficient by using a countercurrent system; however, large volumes of solvent are involved which must subsequently by concentrated. The concentration step introduces losses of extracted sample and increases the mole fractions of the impurities that are present in the solvent. The drawback to headspace analysis is the large volume of helium or nitrogen at high flow rate that is required to extract the heavier, low vapor-pressure pollutants. Grob concluded that it was not possible to purify these gases sufficiently or to effect 100% collection of the extracted material in either a cold trap or on a solid adsorbent.

Grob has overcome these drawbacks in headspace analysis with a closed circuit stripping system which includes a 1.5 to 5 mg carbon disc as the adsorbent to collect the organic materials. The inert-gas stripping system operated at 20 to 30°C and shown in Figure 23 is used to recovery organic solutes up to about C_{14} and a water-vapor stripping system is used for the recovery of higher molecular weight hydrocarbons.

The feature of the system that is especially interesting is the use of the 1.5 to 5 mg charcoal adsorbent disc to collect the organic materials. Charcoal has not been widely used as an adsorbent in pollution studies because of many reported difficulties with irreversible adsorption of some materials and the structural modification of others.[76] Grob has shown that the first problem arises only if the amount of adsorbed material is too low in comparison to the active adsorbent surface. Since the specific surface area of carbon is extremely high, an adsorbing filter with a small amount of carbon should be used. Structural modification of the adsorbates have been avoided by the use of ultrapure carbon, free from organic salts that can catalyze structural changes,[99] and by the use of a solvent instead of heating to desorb the organics.[100] A further advantage achieved by the use of a small amount of adsorbent is that the organics can be extracted from the filter with a small volume of purified solvent.[101] For the 1.5 mg filter, 5 to 15 $\mu\ell$ of carbon disulfide purified by the method of Obach[102] are used and for the 5 mg filter, 10 to 100 $\mu\ell$ of methylene chloride are used. No further concentration of the extracts is required for gc analysis.

The filters are prepared by drawing a weighed amount of carbon into a constricted glass tube and fitting the tube with stainless steel screens to retain the carbon. The filter is washed with dilute nitric acid, water, methanol, and chloroform, and stored in carbon disulfide overnight. After this treatment, the filter can be used without further activation. Between uses, the filters are stored in an organic solvent and appear to have an unlimited lifetime.

The sensitivity of the method is so great that a number of special precautions must be taken. Materials leached out of PTFE have been identified in the chromatogram and the use of this material is avoided in the apparatus. In addition, the filter tube must be held in place by stainless steel or glass sleeves. The apparatus must be absolutely gas tight to prevent the introduction of ambient air. It is noted that the introduction of 1 to 2 ℓ of ambient air into the apparatus would introduce as many pollutants as are present in 5 ℓ of drinking water. Air leakage is avoided by immersing the apparatus in fresh tap water. The method is so sensitive that care must be taken to avoid turbulence when introducing the water sample into the apparatus to avoid dissolution of contaminants from the laboratory atmosphere. The gas used in the stripping circuit must be purified with an auxiliary filter before the analytical filter is installed.

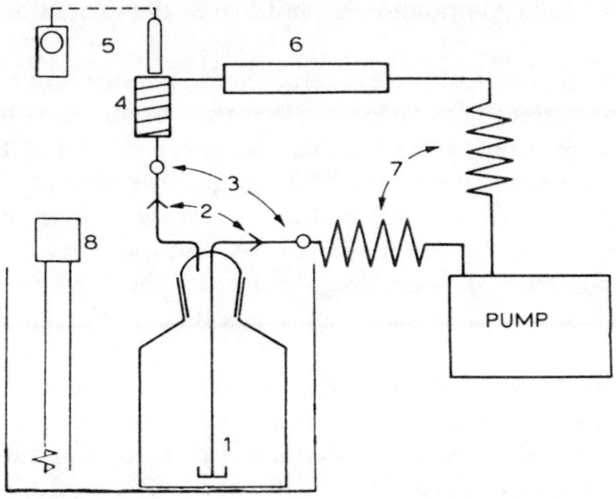

FIGURE 23. Closed circuit apparatus for stripping with inert gas. 1. Glass frit with porosity O; 2. Rotulex bowl joints with PTFE seal (Sovirel); 3. Fused glass-metal connections; 4. Aluminum heating cylinder; 5. Soldering iron (15 W) with time relay; 6. Filter holder; 7. Stainless-steel tubing, 1 m × 3.2 mm o.d., 2.0 mm i.d., 8. Thermostatic water-bath. (From Grob, K. and Zurcher, F., *J. Chromatogr.*, 117, 285, 1976. With permission.)

In drinking water where the concentration of major pollutants is about 5 ppt, stripping with air at room temperature can be used to extract the alkanes up to C_{20} with losses of less than 20%. In lightly polluted water, the limit is C_{22}. For oxygenated, nitrogenated, and aromatic compounds, the recovery factors of the individual compounds should be checked. Stripping with water vapor increases the recovery efficiency for high molecular weight substances and polar materials, but low molecular weight substances are lost. Recovery factors increase with higher concentration of pollutants.

To appreciate more fully the advantages of this extraction procedure, it is valuable to examine the material balance aspects of the system. In a 5-ℓ sample of pure drinking water, 20 ng of an organic pollutant corresponds to a pollution level of 4 ppt. This will represent one of perhaps 50 similar organic compounds in the sample. If the extraction efficiency from water is 50%, the extraction efficiency from charcoal is 100%, and the adsorbates are eluted with 10 $\mu\ell$ of carbon disulfide, the final sample contains 1 ng of pollutant in 1 $\mu\ell$ of solvent. This method, therefore, has the same attraction as thermal desorption (vide infra) in that almost the total extract is used in the analysis. In the case of extremely pure water samples, the 10 $\mu\ell$ extract can be further concentrated.

VII. THERMAL DESORPTION

In thermal desorption, all of the organic pollutants from a water sample that have been adsorbed on a resin column are transferred onto a gc column. This is contrasted to an extraction procedure in which the volume of the extracted sample is reduced to 0.2 to 1.0 mℓ and a 2 to 10 $\mu\ell$ aliquot (or about 1/500) is used for the gc analysis. Thus, by using thermal desorption a gain of a factor of 500 in sensitivity is achieved. The analyst has the option of either utilizing this gain in sensitivity or reducing the volume of the water sample from 1 ℓ to 20 to 25 mℓ. A second advantage of thermal desorption is that volatile compounds are not lost in the extraction and preconcentration steps.

In 1973, Mieure and Dietrich[76] investigated thermal desorption to determine organic pollutants in air and water. They suggested that the samples be collected in an adsorption tube which can also serve as a precolumn for the gc. When the adsorption tube and the analytical column can be heated simultaneously, the configuration in Figure 24 is used. When the adsorption tube and the analytical column must be heated independently the configuration in Figure 25 is used. For the latter configuration the collection column is constructed as an insert to fit into the heated injection port of the chromatograph so that the sample can be flash-heated onto the analytical column with the injection port heater.

Mieure and Dietrich used 1/4" or 3/8" × 4" adsorption tubes filled with 0.2 to 0.5 g of 60 to 80 mesh packing materials that were cleaned and preconditioned by conventional techniques. For the collection of pollutants from air a number of packings were investigated for the adsorption and desorption of specific classes of pollutants. Chromosorb® 101 was useful for acidic and neutral compounds, Chromosorb® 105 was useful for low boiling species and Tenax-GC® was efficient for the collection of basic, neutral, and high boiling substances. For the collection of pollutants from water samples, Chromosorb® 102 was used.

A major difficulty in the analysis of water samples by thermal desorption methods that does not occur in the analysis of air samples is that more than 10 $\mu\ell$ of water remain in the interstitial volume of the collection column packing and this is flashed onto the analytical column with the sample. The presence of this water in the effluent from the gas chromatographic column may necessitate operating at a greater hydrogen flow to avoid blowing out the flame in the detector and the water peak may mask some of the pollutant peaks in the chromatogram.

Versino et al.[103] used Tenax-GC® as the packing in their adsorption tubes. For air sampling a 15 cm × 1 cm glass column filled with 2.5 g of Tenax-GC® was preconditioned at 350°C for 3 hr and then overnight at 200°C with a flow of high purity helium. Two to forty liters of air were sampled by aspirating the air through the adsorption tube at the rate of 0.5 ℓ/min. For water sampling the collection tube contained 4 g of Tenax-GC® mixed with 20% 60 to 80 mesh glass beads. The combination of the packing with glass beads facilitates the flow of water through the column. They sampled 0.5 to 1.5 ℓ of water depending upon the level of contamination. After sampling, the excess water was removed from the column by placing the adsorption tube in a desiccator over phosphorus pentoxide at 10 torr. This drying procedure minimized the losses of compounds that had been collected on the Tenax-GC® and appears to be a much safer way to remove water than drying with a stream of helium.

In the Versino procedure the adsorbed materials were thermally desorbed from the Tenax-GC® at 250 to 270°C with helium flowing at 15 mℓ/min and collected in a glass-lined U-tube. The condensed materials were flash-heated from the U-tube at 300°C into two parallel glass capillary columns with different stationary phases.

Chang and Fritz[104] and Ryan and Fritz[105] have investigated the thermal desorption technique using collection columns packed with XAD resins. The sampling column was an 8 cm × 2 mm i.d. Pyrex® tube that was packed with 120 to 140 mesh XAD-2 or XAD-4 resin. A 2.5 cm resin bed was centered in the tube with silanized glass wool. Connections were made with Swagelok® fittings using PTFE ferrules. The XAD adsorption tube was cleaned prior to use by passing 5 mℓ of distilled water through the tube and then placing the tube in a desorption chamber at 200°C. Helium was passed through the tube at the rate of 20 mℓ/min. The tube was held at 200°C for 4 min and then raised to 240°C for 10 to 15 min. After repeating this procedure four or five times

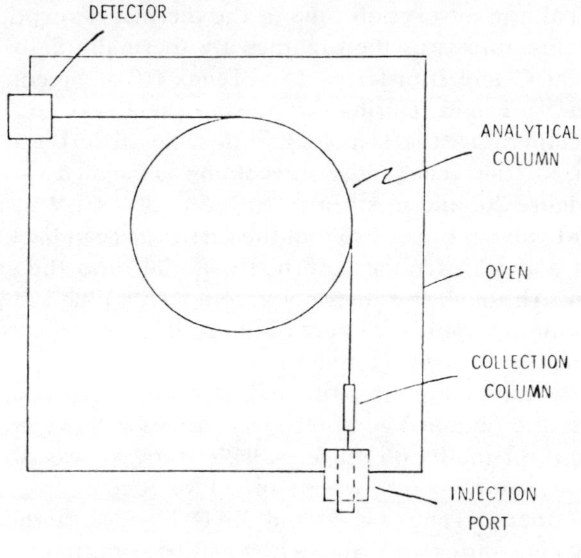

FIGURE 24. Headspace collection column connected to an analytical column in a gc oven. (From Mieure, J. P. and Dietrich, M. W., *J. Chromatogr. Sci.*, 11, 559, 1973. With permission of Preston Publications, Inc.)

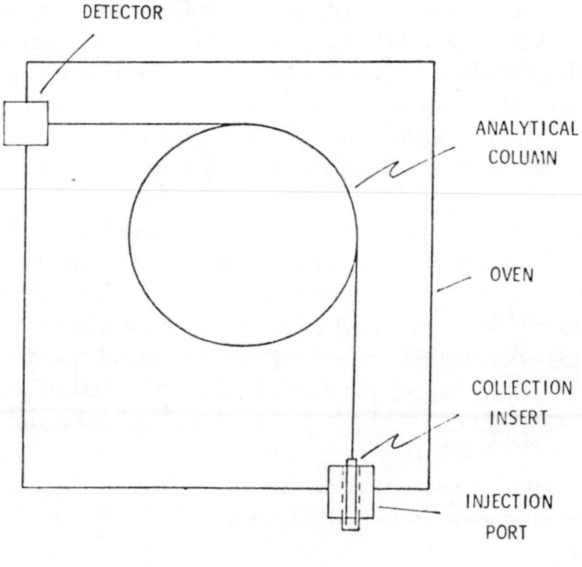

FIGURE 25. Collection column connected into a gc as an injection port insert. (From Mieure, J. P. and Dietrich, M. W., *J. Chromatogr. Sci.*, 11, 559, 1973. With permission of Preston Publications, Inc.)

the tube was capped and ready for use. The water sample was pumped through the collection tube at the rate of 4 to 5 mℓ/min by either a 20 mℓ syringe or a 50 mℓ syringe pump. After sampling, 20 mℓ of air were passed through the tube to remove excess water, then the tube was thermally desorbed immediately or capped and stored. In the procedure of Ryan and Fritz,[105] the thermal desorption is accomplished by connecting the mini-column adsorption tube to the thermal desorption apparatus shown in Figure 26. In this apparatus the organics are thermally desorbed from the XAD resin at 180 to 200°C and transferred to a Tenax-GC® precolumn by a stream of helium flowing at 5 mℓ/min. During the transfer the Tenax-GC® precolumn is held at 45°C; at this temperature the Tenax-GC® does not adsorb water. After the sample has been transferred, the Tenax-GC® precolumn is isolated by means of the 4 port valve shown in Figure 26, and it is heated to 275 to 300°C. When this temperature is reached, the 4-port valve is opened so that the carrier gas can backflush the hot Tenax-GC® precolumn and transfer the desorbed material onto the gas chromatographic analytical column. Although this does not accomplish a plug sample injection, sharp, well-resolved chromatographic peaks are obtained if the gc column is kept cool during this transfer and temperature programming is used.

Ryan and Fritz[105] have tested this thermal desorption procedure with a mixture of model compounds and obtained good recovery factors with water samples spiked with organic solutes in the 1 to 10 ppb range. A 96% recovery was obtained for 1-methylnapthalene when the sampling tube was desorbed for 10 min.

May et al.[90] compared Tenax-GC® and XAD-2 under thermal desorption conditions. Their results are shown in Figure 27. It can be seen that the Tenax-GC® blank gives an acceptable background at 400°C and even the small peaks shown in the figure do not appear at 375°C. The XAD-2 blank is unsatisfactory at 275°C.

VIII. STEAM DISTILLATION

Minsley[106] has reported on the use of steam distillation for the analysis of gasoline, fuel oils, kerosene, and motor oils in water systems. Recently, Veith and Kiwus[107] have developed a modified Nielsen-Kryger® steam distillation apparatus (Ace Glass Co., Vineland, N.J.) shown in Figure 28 for the extraction of pesticides and industrial chemicals from water, sediments, and tissue. The apparatus provides for the simultaneous distillation and extraction of organic solutes from water by a small volume of organic solvent.

A water sample of 1 to 2.5 ℓ or a sediment or tissue sample with added water is placed in a 2 to 3-ℓ round bottom flask; this flask is attached to the apparatus shown in Figure 28 and the water is boiled vigorously for 1 to 7 hr depending on the type of sample. As the water boils the steam distillate rises through the inner tube and condenses on the cooling jacket shown in the figure. The condensate passes through and collects under a layer of solvent whose density is less than that of water and the organic pollutants are extracted into the solvent in this step. Isooctane and toluene are recommended as solvents. The water returns to the distilling flask through the overflow tube which also serves as the solvent trap. At the end of the distillation, the solvent is removed through the stopcock and is either analyzed directly or concentrated prior to analysis.

The recovery of PCBs and related chemicals from 2.5 ℓ of spiked water was evaluated for a 45 to 60-min boiling time. Recovery efficiencies for concentrations in the range of 20 ng/ℓ were between 90 to 100%.

IX. ANALYSIS OF HEAVILY CONTAMINATED ENVIRONMENTAL SAMPLES

The papers discussed up to this point have dealt primarily with the separation and concentration of PAH from model samples, tap water, and raw water. In many environmental samples, especially heavily polluted waters and sludges, the PAH content makes up only a small fraction of the DOC. In addition to the PAH, the sample

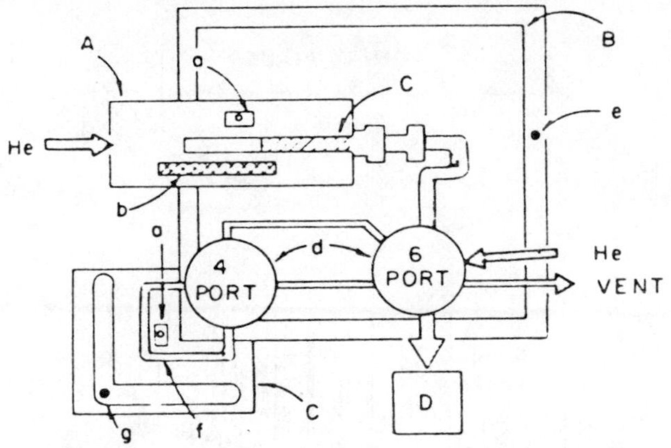

FIGURE 26. Thermal desorption apparatus, (see figure key for letter code).

Key

A: sliding aluminum block
(a) thermocouple
(b) 200-watt cartridge heater controlled by a Variac®
(c) XAD-4 mini-column
B: heated zone insulated with temp-mat glass insulation (Pittsburgh Corning, Pittsburgh, PA)
(d) one four port and one six port zero volume high temperature valve
(e) 200-watt high temperature heating cord controlled by a Variac® (Gears-Col Apparatus Co., Terre Haute, Indiana)
C: removable insulated sheet-metal heating zone (as B) containing the Tenax® pre-column.
(f) Tenax® precolumn (Stainless steel 1/8 in. × 18 cm Tenax GC®, 80-100 mesh)
(g) high temperature heating cord (as e) controlled by a Variac® (Ames lab, Instrument group)
D: gas chromatograph

(From Ryan, J. P. and Fritz, J. S., *J. Chromatogr. Sci.*, 16, 488, 1978. With permission of Preston Publications, Inc.)

contains other hydrocarbons and compounds with oxygen, sulfur, and phosphorus functional groups. In this section a number of studies for isolating the PAH from heavily contaminated water samples will be discussed. Many of these methods incorporate some of the sampling and concentration techniques discussed in the earlier sections.

Pancirov et al.[77] have developed a procedure to separate water pollutants into several classes as outlined in Figure 29. The special feature of the method is the addition of C^{14} labelled B(a)P and B(a)A to the water sample as an internal standard to correct for losses and incomplete and variable recoveries. The radioactive internal standards are added to a 20-ℓ water sample before any separations or concentrations are carried out. The PAH are identified and quantified by injecting an aliquot into a gas chromatograph, trapping each gc peak and measuring its UV spectrum. The measured values are corrected on the basis of the observed recovery of the C^{14} B(a)P and C^{14} B(a)A. The recovery factor of C^{14} B(a)A is used to determine the original concentrations of

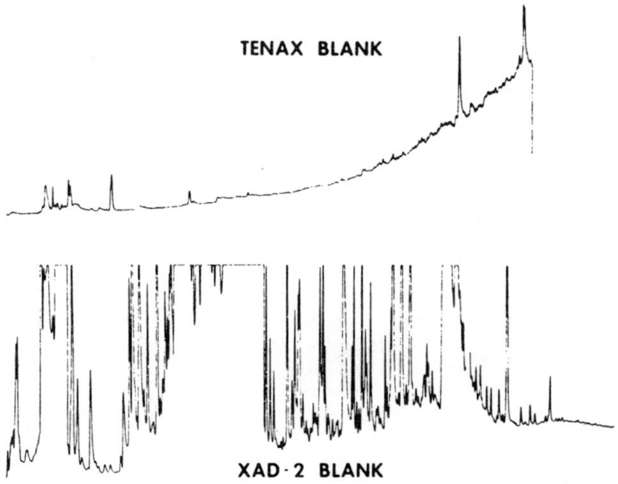

FIGURE 27. Gas chromatograms obtained by flash heating XAD-2 to 275°C and TENAX-GC® to 400°C.

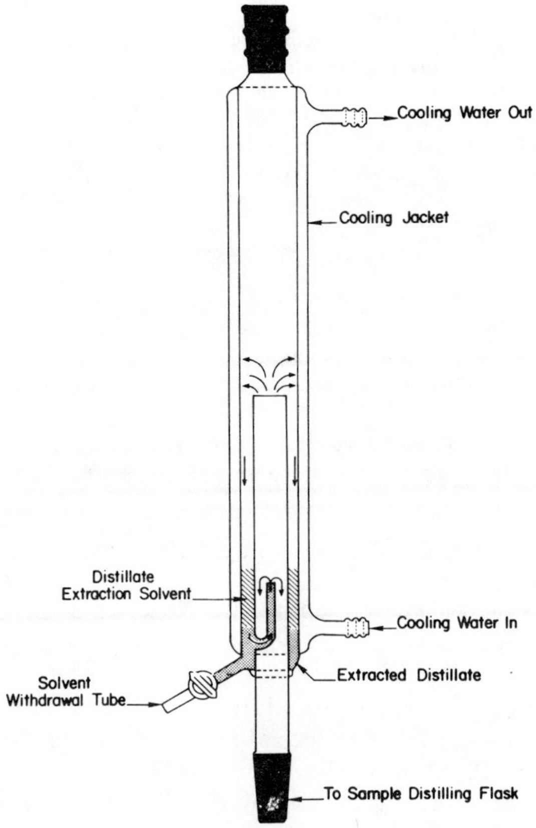

FIGURE 28. Exhaustive steam-distillation and solvent-extraction apparatus.

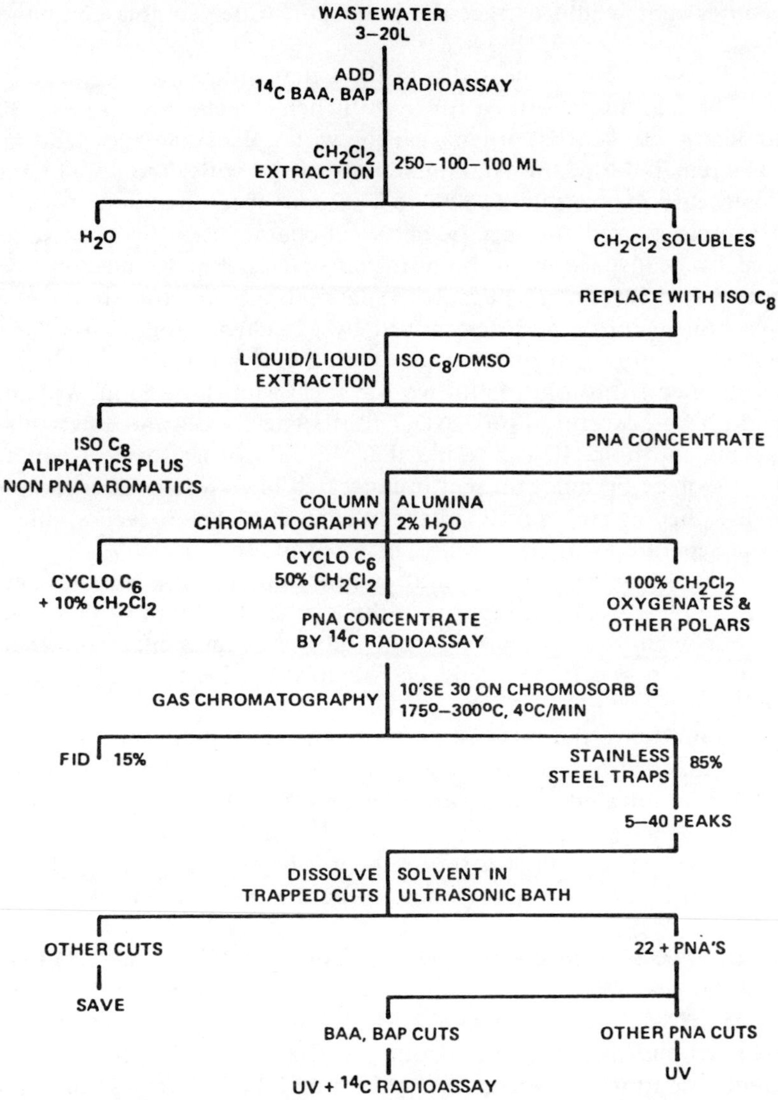

FIGURE 29. GC/UV method for determining PAH in wastewaters illustrating the extraction steps involved. (Reprinted with permission from Lysyj, I. and Russel, E. C., *Symposium on Analytical Chemistry of Petroleum Hydrocarbons in Marine/Aquatic Environment,* American Chemical Society, Washington, D.C., 1978. Copyright by the American Chemical Society.)

B(a)A, chrysene, and triphenylene and the recovery factor of C^{14} B(a)P is used to determine the original concentrations of B(a)P and B(e)P.

Lysyj and Russel[108] have examined the extraction of oils from waste water using the nonpolar solvent Freon® 113 (1,1,2-trichloro-1,2,2-trifluoroethane) as is recommended in Methods 502A and 502B of the *Standard Methods for the Examination of Water and Wastewater.*[74] They found that the Freon® 113 extraction method gave an average recovery of about 56% when used on five different water samples contaminated with crude oil. For this reason, the Freon® 113 extraction was supplemented with procedures that would extract the polar and water soluble components of the crude oil.

Lysyj and Russel[108] have reported a method that utilized a 1-ℓ and 2-ℓ sample of wastewater. The 1-ℓ sample is used for overall chemical characterization; 30 to 100 $\mu\ell$ are used to determine the total organic carbon with a Beckman® or Dohrman® TOC analyzer. The remainder of the 1-ℓ sample is extracted with Freon® 113 and the characteristic hydrocarbon peaks in the 3200 to 2700 cm^{-1} spectral region are examined.

The 2-ℓ sample is used for specific chemical characterization. The volatile solutes are analyzed by headspace analysis, using nitrogen gas and collecting the desorbed solutes on activated carbon. The adsorbed materials are eluted from the activated carbon with carbon disulfide and determined by gas chromatography. The suspended organic matter is removed from the water with a 0.45 μm filter and the dissolved organics are extracted from solution in two ways: (1) with three 50-mℓ volumes of chloroform, and (2) by adsorption on XAD-7 resin. The concentrated extracts were analyzed by gc/ms and hplc. It was found that the chloroform extract contained up to 70% of the dissolved organics present in untreated bilge water. This complex mixture that contained phenols and cresols, nitrogen-containing heterocyclics, alkyl benzenes, and PAH was separated with hplc using a reversed-phase column.

Grimmer et al.[109] reported on a separation procedure (Figure 30) to extract the PAH from sludge samples. It can be seen from the left-hand column of this figure that the PAH with four or more rings make up only 0.01% of the sample and that without a separation procedure the PAH would be swamped by the other pollutants in the gas chromatogram.

The solid samples are freeze-dried, 1,2,7,8-dibenzphenanthrene (10 μg/g) is added as an internal standard, the sample is extracted three times with 50 mℓ of acetone, and the extract is concentrated with a rotary evaporator. The water samples are spiked with the same concentration of internal standard, and 100 mℓ of methanolic potassium hydroxide is added. The sample is refluxed for 2 hr and extracted with three 100-mℓ portions of Freon® 113. After the phases separate the methanol-water phase is discarded.

The acetone extract from the solid sludge sample or the Freon® 113 extract from the water sample (ca. 350 mg) is dissolved in 25 mℓ of cyclohexane and shaken with a solution of 45 mℓ of dimethylformamide (DMF) and 5 mℓ of water. The DMF-water phase is retained and this fraction is diluted to 100 mℓ with water and a further 300 mℓ of cyclohexane to make the liquid composition DMF/water/cyclohexane = 1:2:6. The cyclohexane phase is retained and dried in a rotary evaporator. The residue in the rotary evaporator is dissolved in three sequential 1-mℓ portions of isopropanol and the sample is placed on a Sephadex® LH-20 column and eluted with isopropanol. The first 45 mℓ of eluate contained the two and three ring PAH and are discarded in the Grimmer et al. procedure. The next 135 mℓ of eluate are concentrated and transferred to an alumina column that has been acid treated and contains 10% water. The sample is eluted from the alumina with 120 mℓ of hexane and this eluate contains PAH with four to seven rings. The sample is again concentrated for analysis by gas chromatography. About 20 PAH were identified in a sludge sample using this procedure. The standard deviations of the individual components for five replicate analyses ranged from 1.6 to 11.3%.

Natusch and Tomkins[110] have used a cleanup method in which dimethyl sulfoxide (DMSO) is used to separate the PAH from hydrocarbons and oxygenates in both air and water samples. The method permits the isolation of (1) aliphatic hydrocarbons, (2) alcohols, phenols, and low molecular weight aliphatic and aromatic acids, and (3) PAH, phthalates, aromatic bases, and high molecular weight aliphatic acids.

A pentane extract of the organic solutes is partitioned three times with fresh DMSO.

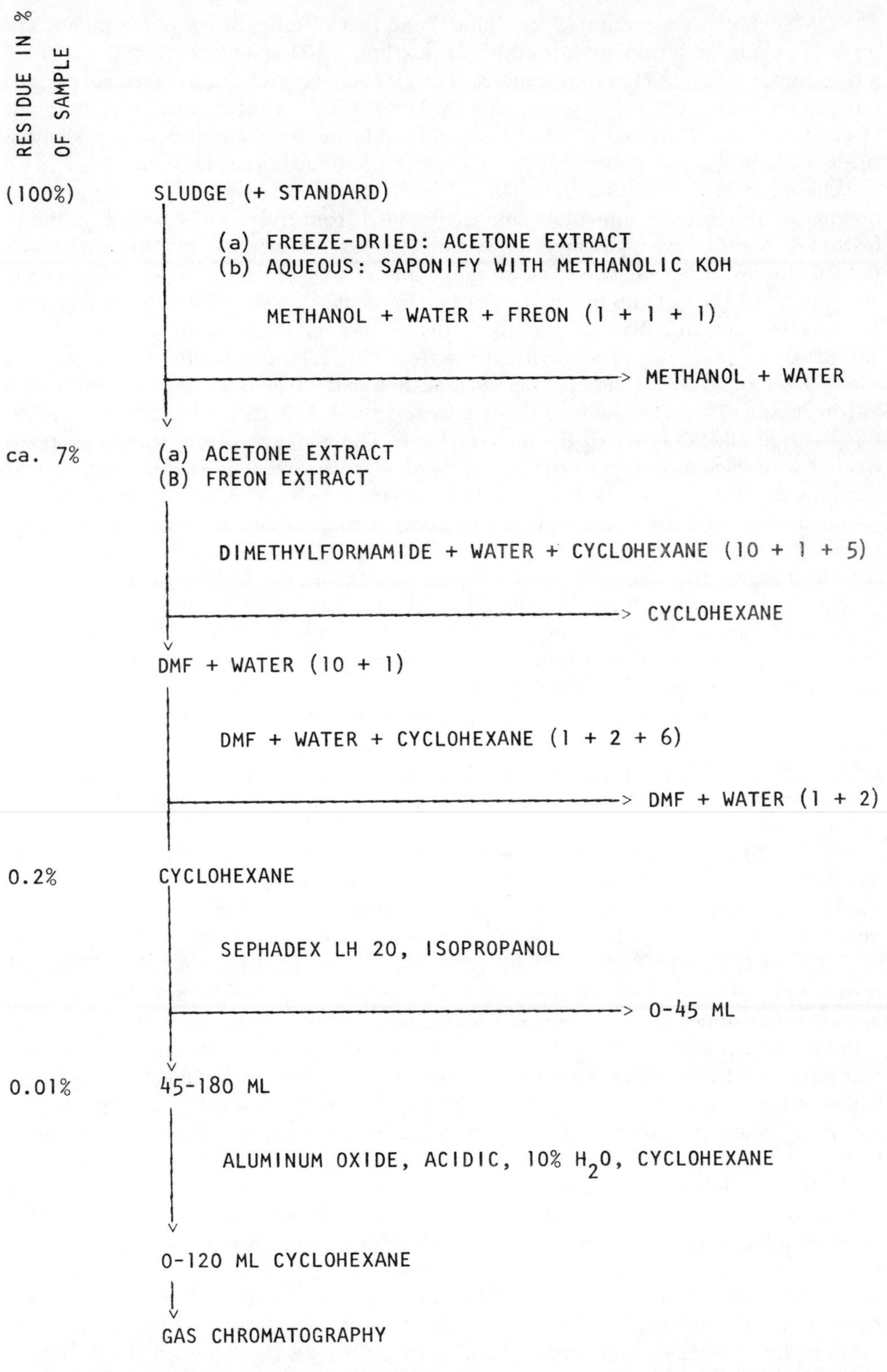

FIGURE 30. Cleanup and extraction procedure for isolating PAH from sewage sludge samples. (From Grimmer, G., Bohnke, H., and Borwitzky, H., *Fresenius' Z. Anal. Chem.*, 289, 91, 1978. With permission.)

The DMSO layers are separated, combined, and two volumes of water are added. The original pentane solution contains the hydrocarbon fraction and the DMSO-water solution contains the PAH and oxygenates. The DMSO-water solution is then partitioned three times with equal volumes of pentane. The PAH are concentrated in the pentane layer in this step. Chromatograms obtained by this method were presented to demonstrate the separations for model systems and extracts of airborne particulates.

Acheson et al.[35] have also used DMSO in an extraction procedure to separate the organic solutes in environmental samples obtained from rivers and sewers adjacent to highways. Nonfluorescing compounds do not seriously interfere with PAH analysis by tlc and fluorescence, whereas they do make analysis by gc very complex. Acheson et al.[35] examined the cleanup procedures of Hoffman and Wynder[111] and Haenni et al.[112] In the Hoffman and Wynder procedure the pollutants are dried to constant weight and dissolved in 40 mℓ of 4:1 methanol-water. This solution is extracted three times with 40 mℓ of cyclohexane and the cyclohexane extract is extracted five times with nitromethane. The nitromethane phase contains the PAH. In the Haenni et al. partition method DMSO replaced the nitromethane. The methanol-water solution was extracted with cyclohexane and the cyclohexane was extracted three times with 15-mℓ portions of DMSO. The DMSO phases were combined and diluted with water and back-extracted with two 25-mℓ portions of cyclohexane. Acheson et al. compared the recovery efficiencies of the two methods for nine PAH for weights of two sets of samples ranging from about 1 μg and 20 μg. In all cases the DMSO partition method gave better recovery and in some cases the recovery was improved by a factor of four. Basu and Saxena have also found the DMSO partition method superior to the nitromethane method for the cleanup of environmental samples collected on open pore polyurethane foam columns (see Table 9).

On the other hand, Sorrell et al.[113] found the DMSO cleanup unsatisfactory for the separation of PAH from other pollutants in their study of the leaching of coatings from water distribution pipes. The DMSO introduced impurities that interfered with the subsequent hplc analysis of the PAH[114] and they selected a cleanup procedure on an alumina column.

Giger and Blumer[115] have described an isolation procedure for PAH from environmental samples that utilizes gel filtration, adsorption chromatography, and charge transfer complexations. They have applied the procedure to the analysis of complex mixtures of aromatic hydrocarbons from marine sediments. A 100- to 150-g sample of sediment is Soxhlet extracted with 275 mℓ of methanol for 24 hr, then 75 mℓ of benzene are added and the extraction is continued for a further 24 hr. The hydrocarbons are partitioned from the benzene methanol mixture with three 75-mℓ portions of pentane, the pentane extract is dried with anhydrous sodium sulfate, and its volume is reduced to 1 mℓ in a rotary evaporator. Elemental sulfur, which would chromatograph with the PAH fraction, is removed by eluting the extract through a copper column with a 1:1 benzene-pentane mixture. The eluate from the copper column is evaporated to dryness, redissolved in a minimum volume of 1:1 benzene methanol, and put on a Sephadex® LH-20 gel permeation column. This column was prepared by placing 20 g of Sephadex® in 1:1 benzene-methanol in a 1.6 cm i.d. tube, giving a packing length of 38 cm. The sample is eluted from the Sephadex® column with benzene-methanol at a flow rate of 6 mℓ/min under slight pressure. The first 50 mℓ of eluate are discarded; the next 50 mℓ contain the PAH and are evaporated to dryness in a rotary evaporator. The aromatic fraction from the Sephadex® is dissolved in 1 mℓ of n-pentane, transferred to a column of alumina over silica gel (both 4 mℓ and deactivated with 3% water), and eluted with n-pentane. The first 20 mℓ of eluate from the column contain the saturated hydrocarbons and most of the olefins and are discarded. The flask which had contained the concentrate from the Sephadex® separation is washed with a 2-mℓ and a 13-mℓ portion of methylene chloride and these washes are used to elute the PAH from the alumina-silica column. A 20 mg sample of 2,4,7-trinitro-9-fluorenone (TNF) is added to the methylene chloride eluate from the alumina-silica gel column to form adducts with the PAH. This solution is evaporated to dryness, the uncomplexed material is dissolved in pentane and discarded. The excess TNF and the adducts are dissolved in methylene chloride and passed through silica gel to split adduct. The PAH are eluted with 75 mℓ of methylene chloride and this volume is reduced to near dryness in a rotary evaporator.

Giger and Blumer point out that for the best separation of TNF and PAH critical attention must be paid to column dimensions, elution volumes and full activation of the silica gel. In their work they obtained a 97% recovery of a 10 μg sample of coronene that was separated from 20 mg of TNF. They also note that the chromatography on the alumina-silica column is an important step. In this separation the lipids remaining in the PAH fraction are removed. Their presence would solubilize the TNF adduct or completely prevent its precipitation. The PAH can now be analyzed by gc/ms or glass capillary gc/ms, or in the procedure of Giger and Blumer this fraction was further separated into seven cuts according to ring-type on an alumina column for UV analysis. Subsequently, Giger and Schaffner[116] have used glass capillary chromatography rather than UV analysis to identify the PAH, and they note that with this analytical procedure the TNF cleanup of the PAH fraction was not necessary.

Sorrell et al.[113] used solvent extraction with cyclohexane followed by a cleanup with an alumina column to isolate the PAH leached from water pipes in preparation for analysis by hplc. The extract was separated into five fractions on an alumina column with (1) pentane, (2) 25% methylene chloride in pentane, (3) 50% methylene chloride in pentane, (4) 50% methylene chloride then 75% methylene chloride in pentane, and (5) methylene chloride.

Recently Keith and Telliard[5a] have reported on the present status of the EPA list of Priority Pollutants and the analytical methods used to monitor them in environmental samples. The list at present contains 129 compounds, 16 of which are PAH. These 16 PAH include the 6 PAH listed by the World Health Organization. The compilation is based on the Toxic Pollutant List and on an EPA report which tabulates all known pollutants in water identified worldwide up to June 1976.[117]

The analytical scheme reported by Keith and Telliard uses three water samples. The first sample is collected in a 40-mℓ vial with a PTFE lined septum, care being taken to exclude air bubbles. Five milliliters of this sample, spiked with bromochloromethane and 1,4-dichlorobutane as internal standards, is used to extract the purgeable compounds by headspace analysis. The purging is carried out with helium in a sparging tube with a fritted glass bottom. The volatile organic compounds stripped from the water are adsorbed in a stainless steel trap packed with Tenax-GC® and silica gel. After the purging, the adsorption trap is backflushed with helium at 180°C and the organics are thermally desorbed to the inlet of a gas chromatography column held at room temperature. Twenty-nine of the Priority Pollutants have been identified in this way.

A second 2-ℓ water sample is used for liquid-liquid extraction. The sample is made strongly alkaline and extracted with methylene chloride. The methylene chloride extract is concentrated in two stages, first with a Kuderna-Danish apparatus and then to 1.0 mℓ with a micro-Snyder distillation column. d_{10}-Anthracene is added to the 1.0 mℓ concentrate as an internal standard. This fraction contains the Base-Neutral Priority Pollutants and 46 of these have been identified by this method. The same water sample is acidified and re-extracted with methylene chloride. The methylene chloride phase is reduced to 1.0 mℓ and d_{10}-anthracene is added as an internal standard. This fraction contains the acid Priority Pollutants and 11 phenols have been identified by this technique.

The third 1-ℓ sample was used to extract the 26 pesticides with a hexane-methylene chloride mixture. The volume of the extract was reduced to 10 mℓ and the components were separated on a Florisil® column. All the determinations were carried out with a gc/ms system.

In the EPA procedure outlined above, the lighter PAH are separated in the headspace step and the higher molecular weight PAH in the neutral/basic fraction of the methylene chloride extraction. We suggest that the use of the procedures outlined may lead to difficulty in determining PAH losses in the concentration step since the internal standard is not added until the volume has been reduced.

Keith and Telliard also list a number of tentative monitoring methods and prime contractors who are studying 12 classes of Priority Pollutants. The analysis of 16 PAH is being investigated at Battelle Memorial Institute. The PAH are extracted from water with methylene chloride and the solvent exchanged with cyclohexane. The extract is fractionated on silica gel and analyzed by hplc using gradient elution with a Perkin-Elmer® HC-ODS reversed phase column. Fluorescence at two different excitation and emission wavelengths is used for identification.

X. SUMMARY AND OUTLOOK

The fundamental problem faced by the analyst is the collection of a representative sample and the identification and quantitation of its components. The additional complications imposed by working with trace amounts are the problems of losses of some sample components and contamination. These constraints favor procedures that involve a minimum of sampling handling. On this basis one should examine critically any preconcentration procedure that involves volume reductions by a large factor. Procedures of this type lead to the loss of volatile components and the concentration of impurities in the solvent. Solid adsorbents are convenient for the sampling of large volumes of water in the field, although this procedure leads to the losses of volatile components. Volatile components are best analyzed in the laboratory by headspace analysis where the sample is collected on a solid adsorbent and transferred to a gas chromatograph by thermal desorption. The analysis of the nonvolatile components can best be carried out by a liquid-liquid extraction method with a minimal volume of solvent that will not require further volume reduction. In all of these procedures, losses should be monitored by the use of appropriately selected internal standards which should be added at the beginning of the analytical procedure. The availability of highly efficient glass capillary gas chromatography/mass spectrometer/data systems and their very large dynamic range should minimize the need for extensive separations that were described in many of the procedures cited in this review. The use of hplc has made it possible to identify and quantify isomeric PAH.

A problem that appears unresolved at present is the PAH that are absorbed on particulate matter. In a number of procedures, the particulate matter is filtered prior to the analysis of the water sample. We think that when filtration is used the sorption of PAH on the filter and on the particulates should be monitored, since in the case of drinking water the particulates are ingested.

It is hoped that rapid advancement of the state of the art will make it possible to establish uniform analytical procedures for the analysis of PAH and other pollutants.

The intense effort by EPA and the many workers in this field should achieve this end in the near future.

Chapter 4

CHROMATOGRAPHIC METHODS OF ANALYSIS FOR PAH

I. INTRODUCTION

A number of analytical techniques have become available for the analysis of PAH in water systems. The concentration of PAH in water systems ranges from the ppt(ng/ℓ) to the ppm(mg/ℓ) level so that a preconcentration step prior to analysis is always required. In samples from heavily polluted waters a cleanup step to separate the PAH from the other organic solutes is necessary. The PAH most commonly encountered in water systems are listed in Table 1. Two approaches are possible: (1) an initial chromatographic separation followed by identification, or (2) a total analysis by spectroscopy. In general the complexity of the sample limits the advantages of the one step spectroscopic approach. The details of spectroscopic analysis as applied to PAH will be discussed later.

We have already reviewed the procedures for sampling, preconcentration, and cleanup of PAH from water systems. Generally the cleanup involves the isolation of the PAH fraction with thin layer chromatography, tlc, or column chromatography.

The problem of separating and analyzing a class of compounds that contains no functional groups and has a large number of structural isomers is a difficult one. All the chromatographic methods have been applied to this problem with varying degrees of success. Cleanup by column chromatography has been used prior to tlc in a number of procedures. Gas chromatography, gc, using a number of supports and stationary phases, glass capillary gas chromatography, $(gc)^2$, and high performance liquid chromatography, hplc, have all been used for PAH analyses.

When the PAH have been separated, a number of methods for detecting and measuring the amount of PAH present are available. Ultraviolet or fluorescence detection are convenient methods subsequent to tlc or hplc separation. Flame ionization detection, or mass spectrometry are generally used with gc or $(gc)^2$. Photoionization and gas phase fluorescence detection methods are being developed.

The factors which must be considered in comparing and evaluating the various analytical techniques are (1) separation efficiency, (2) lower limits of detection, and (3) ease and speed of analysis. Separation efficiency should include complete isolation of the PAH from the other organic pollutants and the resolution of the PAH fraction into its various components. Structural isomers of PAH often vary in carcinogenic activity. For example, benzo(*a*)pyrene, B(a)P, is an extremely active carcinogen whereas benzo(*e*)pyrene, B(e)P, is relatively inactive; dibenz(*a,h*)anthracene is highly carcinogenic while other dibenzanthracenes are not. The lower limits of detection must be sufficient to resolve and identify the PAH isomers. Although time considerations may be significant in some applications, in general they are secondary to good separation and detection. The applicability of the analytical scheme to performance by technicians must also be considered.

II. FRACTIONATION AND CLEANUP PROCEDURES

A. Introduction

The PAH which are found in water can originate from many different sources, e.g., petroleum spills, sewage, industrial wastes, and washout from atmospheric pollution. Generally the PAH are a small fraction of the total dissolved organic carbon which includes aliphatic, olefinic and cyclic hydrocarbons, simple aromatics, organic acids, esters and bases, as well as oxygen, nitrogen, and sulfur heterocycles. A number of methods have been proposed to isolate the PAH and simple PAH derivatives. These methods can often involve solvent extraction and gel filtration techniques which were presented earlier. This chapter deals with column and thin layer chromatographic methods.

B. Thin Layer Chromatographic Methods

Thin layer chromatography separation methods can be carried out relatively simply. Glass plates (20 × 20 cm) are frequently used as supports. Alumina and silica gel are the common stationary phases, although the use of acetylated cellulose has been investigated. The residual water content of the support which can affect the separation efficiency of the adsorbent is a primary concern when using alumina or silica. A number of high temperature activation procedures have been developed to reduce the water content of alumina and silica to a "zero" level. Some workers have found that the addition of 2 to 3% water leads to better fractionation and less tailing. Regardless of the water content of the adsorbent it is important to adopt a constant activation procedure to ensure reproducible results.

The choice of the solvent system to serve as the mobile phase depends on the type of adsorbent and its degree of activation. "Dry" nonpolar solvents such as hexane or pentane are used with completely deactivated as well as partially deactivated alumina and silica. Polar mobile phases such as methanol, acetonitrile, and water are used with nonpolar adsorbents such as acetylated cellulose.

Coating of the stationary phase onto the support is important if reproducible R_F values and quantitative analytical results are to be obtained. Manual and automatic applicators are available for coating the tlc plates. The spotting of the sample is performed with a calibrated microsyringe or micropipet or with an automatic plate spotter. Volumes of 1 to 25 $\mu\ell$ can be delivered with random errors of 1% or less.[118]

The methods of development for cleanup tlc procedures generally involve one-dimensional developments, since only an isolated PAH fraction is desired.

Nielson[119] employed a tlc prefractionation procedure on extracts from gasoline and lubricating oils. The tlc plates (20 × 20 cm, 0.25 mm thick, Macherey-Nagel® SIL G-25 HR) were activated for 1 hr at 100°C. The extract and a standard mixture of PAH were applied as a 6- to 8-mm wide band on the plate. Ascending one-dimensional development in the dark with hexane was followed by development with a 1:1 toluene-cyclohexane mixture. The hexane development was used to minimize possible overloading by a high content of nonvolatile, nonpolar compounds. The PAH fraction was one of six bands identified and was eluted from the silica with two portions of diethyl ether and concentrated to 100 $\mu\ell$ under a dry nitrogen stream.

Lankmayr and Muller[120] employed a similar procedure to isolate the PAH fraction from other organics that were extracted from an environmental dust sample. Their method can be applied to aqueous sample extracts since the PAH studied are also commonly found in water. Commercially available 20 × 20 cm SI 60 tlc plates with 0.25 mm thick layers were used (Merck, Darmstadt, G.F.R.). No activation procedures were given. A 5 cm line from an 800 $\mu\ell$ extract was applied to half of the plate; a test mixture of PAHs consisting of benz(*a*)anthracene, benzo(*a*)pyrene, perylene, benzo(*k*)fluoranthene, dibenz(*a,h*)anthracene, indeno[*1,2,3-cd*]pyrene, and coronene was applied to the other half. Development was carried out with a cyclohexane-benzene (1:5) mixture. All of the PAH were found in a small R_F region between 0.65 and 0.75. Less polar compounds, e.g., aliphatics, were found near R_F values of 1.0 and the polar fraction consisting of nitrogen, oxygen, and sulfur heterocycles was found at low R_F values near the origin.

The aromatic PAH zone was identified under UV light of 254 nm and removed from the plate into a small test tube. The PAHs were eluted from the silica in an ultrasonic bath for 5 min with cyclohexane. Recovery factors obtained for all PAHs were close to the 90% reported for benzo(*a*)pyrene.

Daisey and Leyko[121] isolated a PAH fraction from other organics extracted from airborne particulate matter. A 20 × 20 cm tlc plate coated with 20% acetylated cellulose was spotted 2.0 cm from the bottom at 1.0 cm intervals using a 5.0 $\mu\ell$ disposable capillary pipet with a Wiretrol® dispenser (0.1 to 0.3 mg per spot). A standard mixture of PAH was also spotted on the plate. After development with n-propanol-acetone-water (2:1:1 v/v/v) in a sandwich chamber to a height of 15 cm (approximately 2 hr), the chromatogram was removed, dried, and viewed under a long wavelength UV light. The resulting PAH fraction was spread through an R_F region of ca. 0.15 to 0.80, while the aliphatic hydrocarbons were found at the origin and polar compounds at the solvent front. Figure 31 and Table 16 illustrate these results. The PAH were found in three distinct bands as shown and are removed from the tlc plates by elution with 40% diethyl ether in cyclohexane (by volume). Recoveries from this step are shown in Table 17 and range from 71% for benz(*a*)anthracene to 101% for benzo(*ghi*)perylene. Standard deviations range from 5 to 20% for each PAH studied.

C. Column Chromatographic Methods

Commercial chromatographic columns of various sizes are available; homemade columns from burets, glass, or polyethylene tubing are also used for column chromatography. The same adsorbents and mobile phases used in tlc are also used in column chromatrography. The bands of interest can either be collected as they elute from the column or can be sliced off the column by stopping the mobile phase flow once a desired separation is attained. Since standards cannot be run simultaneously with the sample, standard runs closely duplicating the sample run are required.

Gearing et al.[122] have investigated a column chromatographic technique for isolating the aromatic and aliphatic hydrocarbon classes commonly found in aqueous sediment samples. A standard mixture was studied before application of the method to environmental samples. The standards included n-alkanes with 16, 18, 24, 28, and 32 carbons, pristane, unsaturated paraffins, and aromatics including anthracene and phenanthrene.

Silica gel and neutral alumina (Woelm) were activated by overnight heating at 200°C. Deactivation was accomplished by mixing the stationary phase with an appropriate volume of water. All columns had an adsorbent to sample ratio greater than 100 and were prewashed with hexane before use. The columns were prepared with varying proportions of silica gel under alumina, (v/v), with different degrees of activation. A 3:2, silica gel alumina column, 5% deactivated, and a 4:1 activity grade 1 column had a low fractionation efficiency. A column made up of 2:1 silica gel to alumina (both activity grade 1) was found to be the most efficient.

After the sample was placed on the column, the aliphatic fraction was eluted with two column volumes of hexane and the aromatic fraction was eluted with two column volumes of benzene.

The technique was applied to sediment samples from Narragansett Bay. Fractionation reproducibility was determined by gas chromatographic analysis of the aromatic and aliphatic fractions obtained from replicate column chromatography runs on sediment extracts. The relative standard deviation was 12% for the aliphatic fraction and 27.5% for the aromatics on the basis of 12 runs. Preliminary intralaboratory reproducibility studies gave relative standard deviations of less than or equal to 25%. Gearing et al.[122] found that analogous tlc procedures with identical stationary and mobile phases produced results which were not significantly different.

Table 16
R_f AND R_B VALUES OF SOME POLYCYCLIC AROMATIC HYDROCARBONS SEPARATED BY THIN-LAYER CHROMATOGRAPHY

Compound no.	Compound	R_f	R_B
Band I			
1	Triphenylene	0.16	0.8
2	Benzo[*a*]pyrene	0.23	1.0
3	Anthanthrene	0.31	1.4
4	Chrysene	0.36	1.6
Band II			
5	Benz[*a*]anthracene	0.57	2.5
6	Dibenz[*a,h*]anthracene	0.59	2.6
7	Perylene	0.58	2.5
8	7,12-Dimethylbenz[*a*]anthracene	0.66	2.9
Band III			
9	Benzo[*e*]pyrene	0.70	3.0
10	Benzo[*ghi*]perylene	0.70	3.0
11	Fluoranthene	0.73	3.2
12	Pyrene	0.73	3.2
Polar compounds			
13	Benzanthrone	0.85	3.7
14	Benzo[*h*]quinoline	0.88	3.8
15	Benzo[*f*]quinoline	0.89	3.9
16	Benzo[*c*]cinnoline	0.92	4.0
17	Acridine	0.96	4.2
18	Carbazole	0.97	4.2

Table 17
RECOVERIES OF PAH FROM 20% ACETYLATED CELLULOSE WITH 40% DIETHYL ETHER IN CYCLOHEXANE

Compound	Recovery, %
Benz(*a*)anthracene	71 ± 8
Benzo(*a*)pyrene	90 ± 5
Benzo(*e*)pyrene	72 ± 5
Benzo(*ghi*)perylene	101 ± 22
Chrysene	88 ± 9
Dibenz(*a,h*)anthracene	82 ± 5
Fluoranthene	86 ± 9
Perylene	90 ± 4
Pyrene	89 ± 10

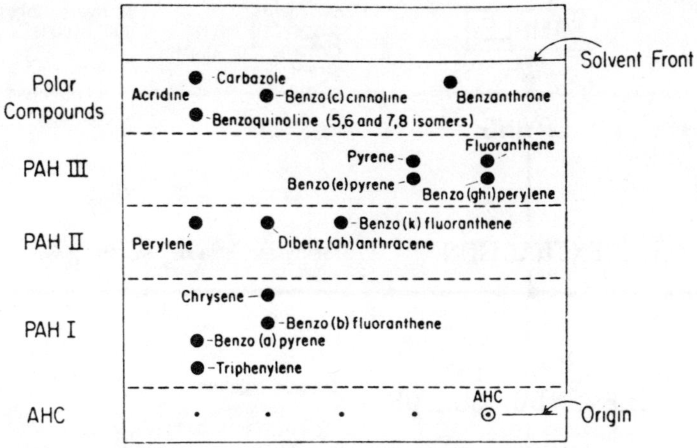

FIGURE 31. Thin-layer chromatogram showing the separation of the PAH, aliphatic hydrocarbons, and other polar aromatic compounds into different regions of the plate. (Reprinted with permission from Daisey, J. M. and Leyko, M. A., *Anal. Chem.*, 51, 24, 1979. Copyright by the American Chemical Society.)

Giger and Schaffner[116] have developed a procedure for isolating PAH from environmental samples utilizing gel permeation and adsorption chromatography (see Figure 32). After the removal of elemental sulfur from the methylene chloride extract by percolation through a column of activated copper, the extracts were carefully evaporated to dryness. The residue was redissolved in 2 mℓ of (1:1) benzene-methanol and subjected to an initial fractionation on Sephadex® LH-20. For this purpose, a 50 × 1.6 cm glass column was packed with a slurry of 20 g of Sephadex® LH-20 in (1:1) benzene-methanol. The same solvent blend was used to elute two 50 mℓ fractions and a third 50 mℓ portion served to flush the column before eventual reuse. The fractionation on Sephadex® LH-20 proceeded by a combination of exclusion and adsorption processes and was favorably characterized by its speed, high column capacity, and the possibility of column reusage. The two resulting 50 mℓ fractions were again carefully evaporated to dryness (rotary evaporator, reduced pressure, ca. 30°C), and redissolved in 1 mℓ of n-pentane. These fractions were applied to silica gel columns (1 cm i.d., 10 mℓ bed volume). Twenty-five milliliters of pentane and 25 mℓ of methylene chloride as mobile phases resulted in two fractions of low and medium polarity, respectively. Flow rates through the column were ca. 2 mℓ/min. The cutpoints of these fractionation steps were established beforehand with naphthalene and coronene standards.

This separation scheme provided two sufficiently pure concentrates: aliphatic and olefinic hydrocarbons in the pentane eluate of the initial Sephadex® fraction and PAH in the methylene chloride fraction of the material which had been more strongly retained on the Sephadex® column, i.e., the latter Sephadex® fraction. This cleanup scheme is an improvement of an earlier procedure developed by Giger and Blumer,[115] discussed earlier.

In 1971, the World Health Organization defined the maximum permissible concentration limit of 200 ng/ℓ for the total concentration of six PAH in drinking water. The six regulated PAH are: fluoranthene, benzo(*b*)fluoranthene, benzo(*k*)fluoranthene, benzo(*a*)pyrene, benzo(*ghi*)perylene, and indeno[*1,2,3-cd*]pyrene. Borneff[123] has devised a tlc analytical procedure for determining the concentrations in water of the six PAH following a cleanup column chromatography step. A small glass column (ca. 6 mm i.d. and 100 mm long) was packed with 0.5 g of alumina, basic, activity II

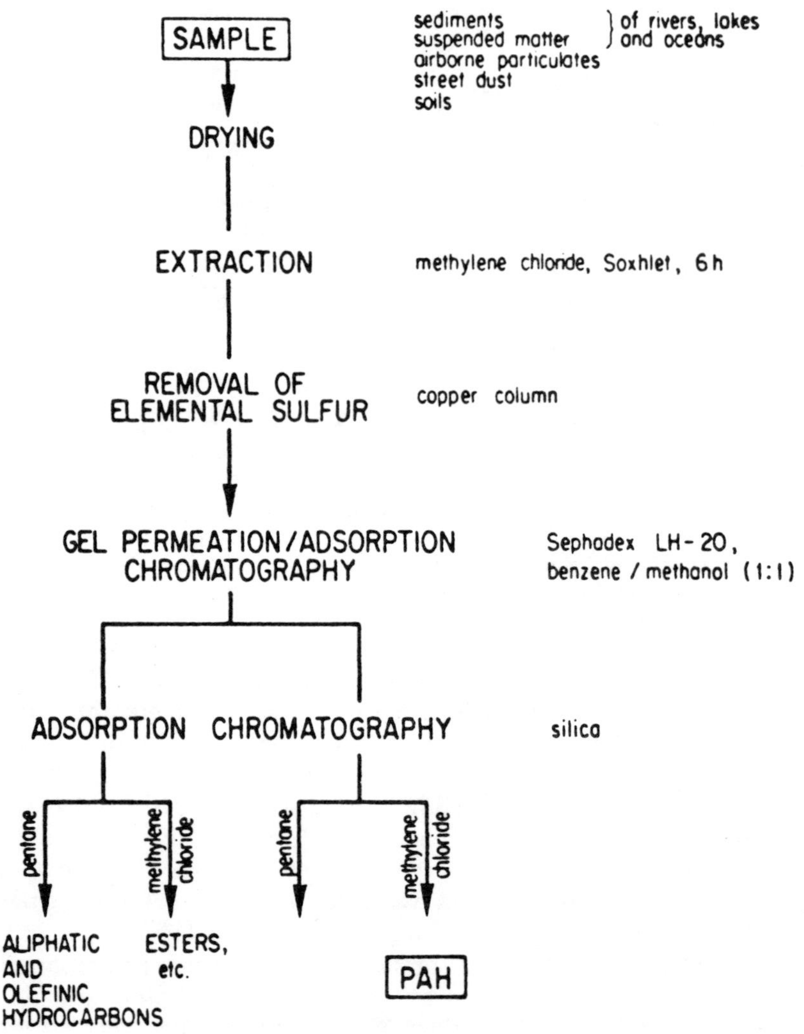

FIGURE 32. Procedure for the isolation of PAH from other components in environmental samples. (Reprinted with permission from Giger, W. and Schaffner, C., *Anal. Chem.*, 50, 243, 1978. Copyright by the American Chemical Society.)

(according to Brockmann). The column was rinsed with 1 mℓ of cyclohexane and then the sample extract was added to the column. The sample extract flask was rinsed with 0.5 mℓ of cyclohexane which was also added to the column. The resulting eluate was discarded. The development was accomplished by addition of 0.5 mℓ of cyclohexane and 3.0 mℓ of a cyclohexane:benzene mixture (1:1). The eluates which contained the PAH were saved for subsequent volume reduction. It was suggested by Borneff that this procedure be tested with a number of reference standards to establish optimum conditions.

Dunn[124] has developed a procedure for determining B(a)P concentrations in marine organisms and sediments which employs an initial column chromatography cleanup step. A column packing of 30 g of Florisil® (60 to 100 mesh) from Matheson Coleman & Bell covered with 60 g of sodium sulfate was prepared in a glass column (4 × 40 cm) with a coarse fritted glass disc. The Florisil® was pretreated by thoroughly washing with distilled water and methanol under suction. Upon the removal of excess methanol, the adsorbent was heated in vacuo at 60°C and then activated at 250°C for 18 hr.

Partial deactivation was carried out after cooling by the addition of 2% water. The column was pretreated with 100 mℓ of isooctane and then the sample which had been extracted into isooctane was added. Two 100-mℓ portions of fresh isooctane were added to elute the aliphatic fraction. The PAH fraction was eluted from the column with three 100-mℓ portions of benzene. The collected eluate can then be reduced in volume. The PAH were extracted next with dimethyl sulfoxide.

Recoveries from the Florisil® column were on the order of 90% or better but depended strongly on the method of adsorbent pretreatment and on the type of sample studied. Relatively little loss of B(a)P from the column occurred during the isooctane elution. Some loss of B(a)P occurred during the benzene elution as a result of tailing. The problem was more severe with samples that had a low organic content or low B(a)P concentration. It was necessary to use batches of stationary phases prepared by the procedure described earlier to obtain consistent results. Fully activated Florisil® contains highly active acidic sites which are capable of leading to chemisorption of PAH such as perylene and was discussed in a paper by Snyder.[125] Such interaction leads to tailing and low recoveries. These sites can be effectively blocked by the addition of 1 to 2% water and Dunn[124] has shown that this procedure led to excellent chromatographic reproducibilities.

Hurtubise et al.[126] have employed column chromatography to isolate the PAH fraction from other organics found in oil shale retort water. The dry column consisted of a 28.0 cm section of polyethylene tubing (0.25 in. i.d., Curtin Matheson Scientific, Inc.) which was loaded to a height of 23 cm with alumina (activity II to III according to Brockman ICN Life Sci. Group). The packing was retained by a glass wool plug which was inserted on the exit end.

A 1.0 mℓ hexane concentrate of the sample was added to the column and developed with 5 mℓ of n-hexane-ether (19:1 v/v). The exact location of the PAH fraction must be ascertained in advance with standards. After the development was complete the purple fluorescent band was located with a UV lamp and the column sliced to give a 5 cm section. The alumina from the sliced section was stirred for 30 min with 20 mℓ of 1,2-dichloroethane, then washed two more times with 10 mℓ dichloroethane portions. The combined extracts were then evaporated to dryness and redissolved in hexane for further tlc separation.

III. THIN LAYER CHROMATOGRAPHY

A. Separation Studies

In addition to its effectiveness as a cleanup technique for isolation of the PAH fraction, tlc has often been used as the analytical technique for the qualitative and quantitative determination of individual PAH concentrations. The same adsorbents described earlier, silica, alumina, and acetylated cellulose, or combinations of these are commonly employed as the stationary phase. The solvents mentioned earlier are also used as mobile phases.

Spotting procedures for tlc were discussed earlier. Development procedures are more sophisticated for PAH separation and identification than for cleanup procedures. The spotted plate is dried either by a stream of air or by equilibration in an inert atmosphere. Glass development chambers are most often used although specialized chambers with controlled temperatures and humidities are available. The simplest possible technique is multiple development with a single solvent system. The development is considered complete at the point at which a suitable resolution of PAH has been effected. A variant of this method employs different solvent systems before each successive development. Multiple or stepwise development in two dimensions is also a method by which successful PAH separations can be effected. In this procedure, the sample to be analyzed is spotted in one corner of the plate and then allowed to develop in one dimension to a fixed distance. The plate is then removed from the developing chamber and dried. Upon rotation of the plate by 90°, the plate is developed with the same or different solvent system in the new direction. Resolution of PAH with this technique is excellent.

Borneff[123] devised a two dimensional tlc technique for separating the six World Health Organization PAH after an initial column chromatography cleanup. A mixed adsorbent stationary phase consisting of aluminum oxide and 40% acetylated cellulose in a 7:3 weight ratio was used. For the preparation of five tlc plates, 28 g of aluminum oxide was thoroughly mixed with 12 g of 40% acetylated cellulose and 65 g of ethanol. After the plates had been coated, they were activated for 30 min at 130°C in a drying oven. The sample extract was applied as one spot in one corner of the tlc plate about 1.5 cm from the two edges. The first development took place for 30 min with a mixture of n-hexane-benzene, (9:1 v/v). After drying and rotation of the plate by 90°, development proceeded for 60 min with a mixture of methanol-ether-water (4:4:1 v/v/v). Throughout the entire two-dimensional development, the tlc plate was kept in the dark. Since standards could not be spotted simultaneously with the sample, standard runs were carried out under identical conditions. The six World Health Organization PAH were clearly resolved under these analytical conditions.

Pierce and Katz[127] and Katz et al.[128] developed a two stage tlc procedure for separating PAH extracted from airborne particulate matter. Though most of the PAH studied were of molecular weight 276 and above, (compounds unlikely to be found to a large extent in water), this technique was also applied to the benzopyrenes and benzofluoranthenes and is applicable to PAH of any molecular weight.

The analytical scheme was pretested on a standard solution of PAH, concentrations of approximately 0.1 mg/mℓ in toluene. In the first stage, the separation was carried out on a 20 × 20 cm tlc plate coated with neutral aluminum oxide. A 2.0 $\mu\ell$ aliquot of standard solution was spotted 1.5 cm from the bottom of the plate, along with individual PAH standards. Pentane-ether (19:1 v/v), was the mobile phase. Development proceeded for 13 to 15 cm, (20 to 30 min), in a sandwich chamber. This procedure provided a class separation of the 13 component PAH mixture into four groups. Each group was removed from the plate and the PAH were eluted with 20 to 30 mℓ of hot dichloromethane for 15 min. The extract was evaporated to dryness with dry nitrogen and was redissolved in toluene. Junk et al.[41] have pointed out that 10 to 80% of all but the extremely low vapor pressure solutes were lost using a free evaporation step of this sort. However, Katz only studied higher molecular weight PAH and obtained recoveries ranging from 80 to 95%.

The PAH in the four fractions were further resolved in the second stage of the tlc procedure. In this step, an 0.25-mm thick stationary phase of 40% acetylated cellulose was employed. Spotting and developing was carried out in the same manner as in the first stage. The mobile phase was n-propanol to acetone to water (2:1:1 v/v). Table 18 lists R_F values for the PAH relative to benzo(*a*)pyrene. It can be clearly seen that the benzopyrene and benzofluoranthene isomers were resolved. The general applicability of the procedure is limited by the evaporative step and the number of manual steps. If the volume reduction were to be performed with a micro Kuderna-Danish apparatus the procedure should work well with PAH of lower molecular weight. The procedure was tested using two dimensional tlc with a mixed aluminum oxide-cellulose acetate stationary phase but the inability to apply an internal standard to the same plate precluded its use for quantitative analysis. Qualitative analyses give excellent results with the two-dimensional scheme. In Borneff's work,[123] standards were applied to identical tlc plates and not to the same plate as the sample mixture. A comparison

Table 18
RESOLUTION OF SEVERAL PAH BY TWO STEP TLC TECHNIQUE

Compound	1st Step: class separation aluminum oxide, pentane-ether (19:1 v/v) $R_{B(a)P}$[a]	2nd Step: 40% acetylated cellulose, n-propanol-water-acetone (2:1:1 v/v/v) $R_{B(a)P}$
Benzo(*a*)pyrene	1.00	1.00
Benzo(*e*)pyrene	0.98	6.61
Benzo(*j*)fluoranthene	0.99	1.51
Benzo(*k*)fluoranthene	1.00	3.07
Perylene	0.94	4.52
Benzo(*ghi*)perylene	0.89	6.78
Pyrene	1.18	6.40
Dibenz(*a,h*)anthracene	0.78	4.53

[a] $R_{B(a)P}$ refers to relative R_F values with R_F of B(a)P equal to 1.00.

Reprinted with permission from Katz, M., Sakuma, T., and Ho, A., *Environ. Sci. Technol.*, 12, 909, 1978. Copyright by the American Chemical Society.

of relative deviations resulting from the use of external versus internal standards is not available.

Mainwaring and McGuirk[129] have reviewed all of the basic tlc approaches for PAH analyses and have developed a simplified two-dimensional tlc scheme based on the work of Pierce and Katz.[127] A mixed stationary phase, 20% acetylated cellulose:alumina (1:3) was activated at 100°C and 50 $\mu\ell$ samples were developed. The development in the first direction was performed with a pentaneher (19:1 v/v) mobile phase. The development in the orthogonal direction was carried out with ethanol:methylene chloride:water (20:10:1 v/v/v). The results indicated that the benzo(*a*)pyrene moved to near the top of the plate after the first development. However, in the orthogonal second development, the benzo(*a*)pyrene spot travelled behind all of the other spots on the chromatogram. Therefore, to take advantage of the high mobility of B(a)P in the first direction and ensuing small relative shift in the second direction, Mainwaring and McGuirk developed a reversed two dimensional technique, i.e., the second development is in a 180° opposite direction from the first. The usefulness is apparent. The B(a)P spot which is near the top in the first development will be further resolved from the other spots in the second development as a result of lagging behind the other spots. Clearly, a third development could again be made in the opposite direction with the pentane to ether mobile phase. A major advantage of this approach is the ability to spot as many as six to eight samples in a row on the bottom of the plate. Standards or multiple samples can be simultaneously run.

A number of other studies using a tlc separation step for PAH have been published in addition to the work cited above. Bories[130] used polyamide as a stationary phase for the separation of a number of PAH such as B(a)P, benzo(*k*)fluoranthene, perylene and coronene with a mixture of toluene, and methanol (4:1 v/v) as the mobile phase. Shiraishi et al.[131] studied the separation of 28 PAH on Kieselguhr G impregnated with 10% liquid paraffin with a 13:7 acetonitrile-water solvent mixture as the mobile phase. The separations obtained were poor compared to those obtained with the silica/alumina/acetylated cellulose substrate discussed earlier.

B. Methods of Detection

Qualitative analysis of the separated tlc spots can be easily effected by the use of a long wavelength UV lamp. PAH can be identified by comparison of the R_F values to standards. Prior knowledge of the standard concentration even allows a visual semiquantitative analysis of unknown concentration through comparison of spot size. Borneff[123] has noted that reference chromatograms can be prepared which if sprayed with a nonfluorescing preserving dispersion are usable for a month or more as a semiquantitative and qualitative visual standard if exposure to light is minimized.

Quantitative analyses of the PAH are generally performed by one of two different methods: (1) measurement of the solution fluorescence after elution of the PAH spots from the tlc plate, and (2) measurement of the fluorescence intensity on the tlc plate.

In the first method, the individual PAH can be qualitatively identified on the thin layer plate, scraped off the plate, and eluted with an organic solvent. It is important that this step result in quantitative recovery. Automatic spot scrapers and elution devices are available which can process multiple samples with speed, accuracy, and precision. The solution of PAH is diluted to a specified volume, if necessary, and the fluorescence emission is measured at a predetermined emission wavelength. In some cases, the eluting solvent is unsatisfactory for followup fluorescence analysis due to a high level of a self-fluorescence. In such cases, the eluting solvent is usually evaporated with a stream of nitrogen and the resultant residue is redissolved in a more suitable solvent. Care must be taken to guard against losses of volatile PAH in the evaporation step. With the use of standards and prepared calibration curves, unknown concentrations can be determined.

In the second method, the fluorescence can be measured directly on the tlc plate. The excitation radiation is directed through slits onto the tlc plate, and the resulting fluorescence can be measured with a densitometer. The tlc plate is slowly moved past the slit at a constant rate and the fluorescence intensity is recorded on chart paper. The area of the resulting peak is proportional to fluorescence intensity which in turn is related to PAH concentration. Manual or electronic integration techniques can be used in the area measurements.

Katz et al.[128] and Pierce and Katz[127] determined the fluorescence intensities of PAH separated by the tlc methods discussed earlier by eluting the spots from the plates with diethyl ether. These solutions were deoxygenated with a dry nitrogen stream and the solution fluorescences were measured. Quantitative results were obtained by comparison with standards. Scholz and Altmann[132] measured B(a)P in ground water by fluorescence analysis after prior tlc separation on silica gel. After locating and scraping off the B(a)P spot from the plate, the B(a)P was eluted with cyclohexane. Upon evaporation of the cyclohexane to dryness and dissolution of the residue in dioxane, the B(a)P in solution was excited at 365 nm and the emission measured at 429 nm. A concentration of 0.1 ng/ℓ was detectable with relative errors of ± 15%.

Dunn[124] measured B(a)P in hexadecane solution fluorimetrically after removal of the PAH from the tlc plate. A baseline was drawn between the minima in the fluorescence spectrum at 418 and 448 nm and the peak height at 430 nm was determined above this baseline. The sensitivity of the method was 0.1 μg/kg with a precision of 6%.

Basu and Saxena[34] have determined the fluorescence intensities of the six World Health Organization PAH after prior tlc separation. Direct measurement of fluorescence on the plates at room temperature was performed with an Aminco Bowman® thin-film scanner attached to a spectrophotofluorimeter. The measurement was carried out in this way in order to avoid losses of PAH during removal of the spots from the tlc plates and to eliminate errors from the introduction of stray fluorescence resulting from solvent impurities. Table 19 lists the excitation and emission wavelengths used for the six PAH and the limits of detection. For all measurements, a quinine sulfate standard was used to correct for light source variations.

Table 19
DETECTION LIMITS OF THE SIX WORLD HEALTH ORGANIZATION REGULATED PAH WITH A TLC-FLUORESCENCE METHOD

Compound	Excitation wavelength (nm)	Emission wavelength (nm)	Absolute limit of detection (ng)	Limit in 60 liters of water (ng/ℓ)
Fluoranthene	365	458	140.0	2.3
Benzo(*j*)fluoranthene	365	427	7.5	0.1
Benzo(*k*)fluoranthene	365	428	5.0	0.1
Benzo(*a*)pyrene	365	427	10.0	0.2
Indeno(*1,2,3-cd*)pyrene	365	467	10.0	0.2
Benzo(*ghi*)perylene	365	416	20.0	0.3

Reprinted with permission from Basu, D. K. and Saxena, J., *Environ. Sci. Technol.*, 12, 791, 1978. Copyright by the American Chemical Society.

Woidich et al.[133] were able to quantitatively estimate concentrations of PAH after tlc separation by in situ fluorescence measurements. Plots of peak area versus concentration of B(a)P in picograms were linear from 50 to 10,000 pg, and the limit of detection was 10 pg. The fluorescence measuring conditions for 14 other PAH were listed. Kaschari and Reiter[134] measured the fluorescence intensities of B(a)P, B(k)F, IP, B(b)F, and perylene in oil distillates and refinery wastewaters, after tlc separation on acetylated cellulose. They were able to detect between 0.1 and 2 ng of each PAH.

Hurtubise et al.[126] measured the fluorescence intensity of B(a)P directly on a tlc plate. A Kontes® densitometer was used and the recorded peak heights, corresponding to the fluorescence intensity of the standards, were plotted versus the amount of B(a)P in nanograms. Unknown concentrations of B(a)P could then be directly determined from the calibration curve. The limit of detection was found to be 0.9 ppb. With larger sample sizes spotted onto the tlc plate, this limit was lowered to 0.08 ppb.

C. Future Trends

The overwhelming advantages of a tlc technique for the analysis of PAH arise from the simplicity of the operation and the inexpensive experimental requirements. A quick screening of the PAH fraction can be performed to determine whether followup gc, hplc, or other techniques are necessary. The separation efficiency of tlc is high for isomeric PAH. Fluorescence detection of the resolved spots allows subnanogram levels of PAH to be determined. It is our conclusion that a backup tlc analysis system should be considered by all investigators.

IV. HIGH PERFORMANCE LIQUID CHROMATOGRAPHY

A. Introduction

The application of high performance liquid chromatography, hplc, to the analyses of PAH in environmental samples has increased significantly in the last few years. Two factors are primarily responsible for the utilization of hplc in environmental analytical laboratories. The first factor is the development of chemically bonded stationary phases for hplc which have greatly improved the separation of PAH isomers. For example, chrysene/benz(*a*)anthracene/triphenylene are baseline resolved with a C_{18} column packing. The highly successful separations achieved with this packing are reflected by the fact that all the papers which reported on the use of hplc at the Fourth International Symposium on PAH in Columbus, Ohio, October 1979, utilized C_{18} stationary phases in the separation system.[135] Although performance variations exist among the C_{18} packings available from various manufacturers, it is clear from personal discussion at the symposium that the performance of C_{18} packed columns for PAH analyses is unsurpassed at present.

The second factor is the development of highly sensitive fluorescence detectors which have extended the detection limits of PAH to picogram and even subpicogram levels. A useful discrimination is provided by this type of detector since almost all PAH are capable of fluorescing whereas many other organic pollutants are nonfluorescing. With the use of a fluorescence detector system, both an excitation and an emission wavelength can be selected and this provides another degree of selectivity.

A critical evaluation of these and several other developments in PAH analyses are presented.

B. Separation Studies

1. Introduction

Hplc can be viewed as a logical extension of classical column chromatography. An essential difference between column chromatography and hplc is the size of the column packing materials. In column chromatography the particle size is large enough to permit the mobile phase to flow by gravity alone; in hplc the microparticulate packing material, (5 to 10 μm), requires the use of a high pressure pumping and injection system. The high surface to volume ratio of the microparticulate packing results in an hplc separation efficiency comparable to that obtained by gas chromatography.

The efficiency of hplc separations is influenced by a number of factors: column dimension, sample volume, pressure, mobile phase flow rate, and the composition of the stationary and mobile phases. This work will consider specifically the selection of the column packing and an appropriate mobile phase system best suited to that packing for the separation of PAH.

Historically, the first types of packings investigated were the traditional column chromatography materials, e.g., silica and alumina. For these polar adsorbents, nonpolar mobile phases such as hexane, heptane, pentane, benzene, toluene, and chloroform were employed. Separation efficiencies of PAH with these systems were only fair. The degree of separation was determined by selective adsorption on the silica or alumina surface. Attempts to improve the separation by coating the silica or alumina with various substances had limited success because of difficulties in achieving homogeneous surface coverage. Column bleed presented additional problems. The primary requirements of the mobile phase for the silica and alumina packings were the nonpolarity and the "dryness" of the solvent. A number of n-alkanes or mixture of n-alkanes with benzene, toluene, or diethyl ether have been used.

With the development of chemically bonded stationary phases many of the problems of the traditional chromatographic systems are reduced. Stationary phase bleed is minimized with the chemical bonding of the stationary phase to the support. Furthermore, the nature of the silica surface is altered by the bonded nonpolar groups such as the octadecyl group, so that the need for scrupulously dry solvents to prevent the deactivation of the silica surface no longer exists. The technique utilizing a nonpolar stationary phase and a polar mobile phase is called "reversed phase" hplc, as opposed to normal phase hplc with polar adsorbent packings. The mechanism of separation of PAH in reversed phase hplc mainly involves the differential solubility of the PAH in the mobile phase although interactions with the C_{18} group or unreacted support surface may also affect the retention mechanism.

Since the mechanism of separation in hplc depends on the differential solubilities of the PAH in the mobile phase, the proper choice of solvents is critical. Generally, the PAH are eluted with a mixed solvent system rather than isocratically, i.e., with a single solvent. The term "gradient elution" is applied when the composition of the mixed solvent is changed linearly from a water-rich to an organic-rich mixture during the course of the run.

Snyder et al.[136] estimate that nearly 70% of all hplc separations employ gradient elution. A similar solution exists in the field of gas chromatography where temperature programming predominates over isothermal gc separations. The following advantages for gradient elution separations employing a linear solvent strength program are:[136] (1) regular band spacing throughout the chromatogram, (2) constant band widths in the chromatogram, for all bands, (3) comparable resolution or effective plate number for early and late eluting bands, and (4) conceptual simplicity which permits the prediction of separations with variation of experimental parameters. Optimum conditions for a gradient elution program depend on a knowledge of several instrumental parameters. The organic solvent to be used with water depends on the compounds to be separated; acetonitrile and methanol are the most commonly used for PAH. The compositions generally reported range from 20 to 50% organic solvent to 100% organic solvent. The gradient steepness, i.e., the percentage increase in organic solvent, depends upon the particular solvent and the column dead time. For a 1 min dead time with acetonitrile, a 6.7% gradient is optimal for maximum resolution. An excellent discussion on the "fine-tuning" aspects involved in the design of a solvent program can be found in Dolan et al.[137]

2. Reversed Phase High Performance Liquid Chromatographic Separations of PAH — C_{18} Bonded Phases

Grushka and Kitka[138] recently reviewed the use of chemically bonded stationary phases in hplc. Although most of the reported hplc studies have been carried out with commercially available C_{18} bonded phases, it is informative to understand how the bonded phases are prepared. At present, silica gel is the most commonly employed support surface. This surface has a fixed number of hydroxyl groups which can be reacted with mono, di, or trichlorooctadecylsilanes (see Figure 33). The first step involves heating the silica surface under vacuum to 150°C to remove bonded water. To prepare the C_{18} bonded stationary phase, trichlorooctadecylsilane is reacted with the silica gel (see Figure 33C). For reactions with dry solvents a monomeric (one C_{18} group) stationary phase should result. If there are traces of protic impurities in the solvent, the free chloro group can be hydrolyzed with the resultant hydroxyl group capable of further reaction. This leads to a polymeric stationary phase as seen in Figure 33C. The free hydroxyl groups left after the bonding of the C_{18} group can be removed by silanization. The extent of polymeric coating can have an effect on separations.[138] The extent of surface coverage can be calculated with details found in Reference 138.

Wise et al.[139] recently compared the performance of several commercially available C_{18} columns. The column packings were: LiChrosorb® RP-18, Partisil®-5 ODS, Dupont Zorbax® ODS, MicroPak® CH-10, Nucleosil® 10 C_{18}, (Vydac® 201 TP, essentially identical to Perkin-Elmer HC-ODS), Radial Pak® A, and MicroPak® MCH-10. Studies were performed on the separation of PAH standards using a mobile phase of acetonitrile and water with a gradient ranging from 70 to 100% acetonitrile over 40 min. They found that only the Vydac® column gave a baseline separation of chrysene and benz(*a*)anthracene. Only two column packings were considered satisfactory; the Dupont Zorbax® ODS had the highest efficiency, i.e., gave the sharpest peaks, and the Vydac® 201 TP exhibited the best selectivity towards PAH. The polymeric Zorbax® column is useful for complex or "dirty" samples where a high efficiency is advantageous and the Vydac® column is useful for the resolution of individual PAH.

The work being conducted at Perkin-Elmer by Ogan et al.[140,141] is at the forefront of the studies on PAH separations. Extracts of various aqueous samples have been analyzed by hplc coupled with fluorescence detection. The extracts were obtained by liquid-liquid extraction of 1-ℓ samples with dichloromethane followed by Kuderna-Danish concentration. A model solution of 16 PAH including the 15 on the EPA Priority Pollutant List was also studied.

The chromatographic conditions for the analys s are shown in Table 20. The solvent program depicted in Figure 34 was employed. Equilibration of the column between runs was carried out by passing a 20% acetonitrile mobile phase through the column for 6 min. This was followed by 50% acetonitrile for 15 min. The sample loop was rinsed with 30 $\mu\ell$ of acetonitrile and air dried before the introduction of a 15 $\mu\ell$ sample of extract into the loop.

Figure 35 illustrates an example of the PAH separation by hplc that they obtained with the use of isocratic elution, (80% acetonitrile in water), at room temperature. It can be seen that anthracene and phenanthrene are only partially resolved and peaks #14 to 16 are beginning to broaden significantly. Figure 36 illustrates the separation of the 16 PAH achieved by Ogan's group with the solvent program of Figure 34. Dibenz(*a,h*)anthracene and benzo(*ghi*)perylene were only partially resolved. Complete baseline resolution was achieved for all of the other isomers that are generally difficult

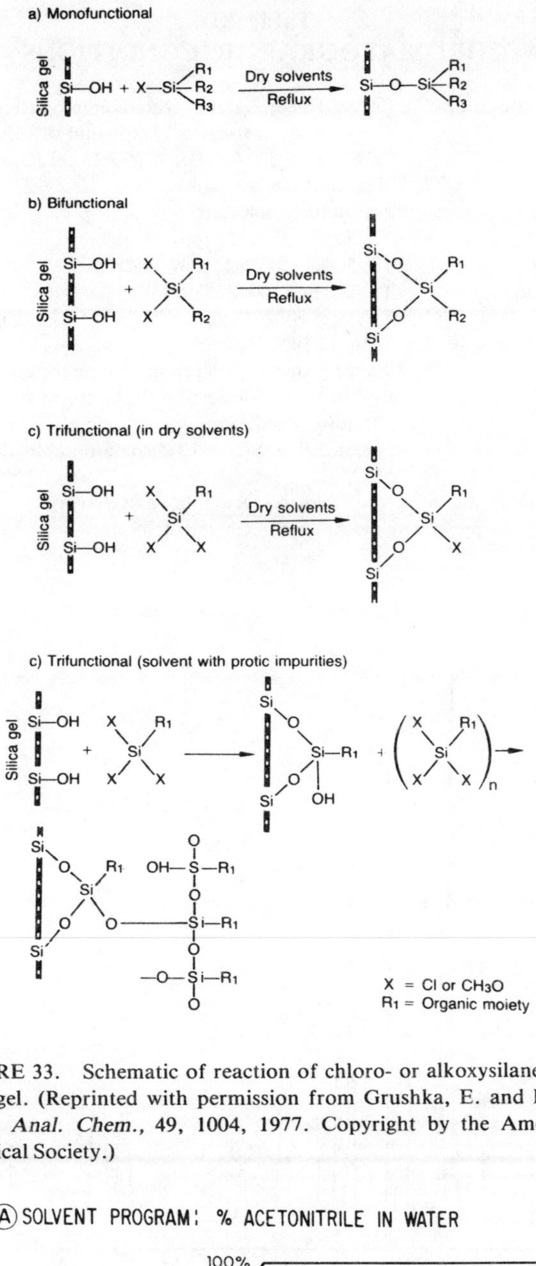

FIGURE 33. Schematic of reaction of chloro- or alkoxysilanes and silica gel. (Reprinted with permission from Grushka, E. and Kitka, E. J., *Anal. Chem.*, 49, 1004, 1977. Copyright by the American Chemical Society.)

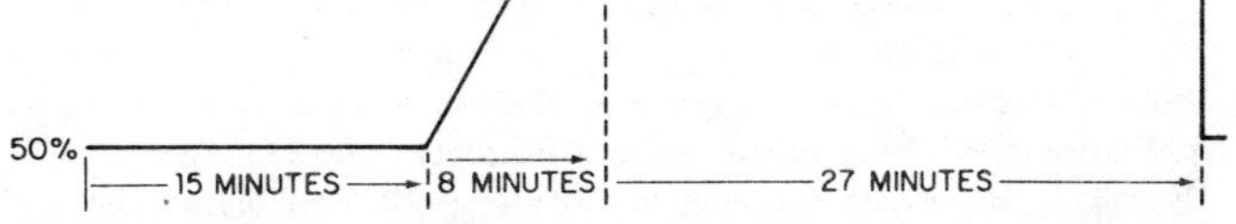

FIGURE 34. Solvent program for chromatographic analysis showing the pump control profile. (Reprinted with permission from Ogan, K., Katz, E., and Slavin, W., *Anal. Chem.*, 51, 1315, 1979. Copyright by the American Chemical Society.)

Table 20
CHROMATOGRAPHIC CONDITIONS

Liquid chromatograph	Perkin-Elmer Series 3, with dynamic stirrer accessory, Rheodyne® Model 7120 injection valve, 10 *μl* loop.
Column	Perkin-Elmer HC-ODS, 0.26 × 25 cm 10 μm, C_{18} packing, part no. 089—0716.
Mobile phase	Acetonitrile in water: T1: 50%—50%, 15 min T2: 50%—100%, 8 min, linear T3: 100%—100%, 27 min T Purge: 20%, 5 min T Equil: 50%, 15 min 0.5 m*l*/min flow rate room temperature.
Detector	Perkin-Elmer Model 650-10LC, spectrofluorimeter. Bandpass = 12 nm.
Data collection	Perkin-Elmer Sigma 10 chromatography data station.

Reprinted with permission from Ogan, K., Katz, E., and Slavin, W., *Anal. Chem.*, 51, 1315, 1979. Copyright by the American Chemical Society.

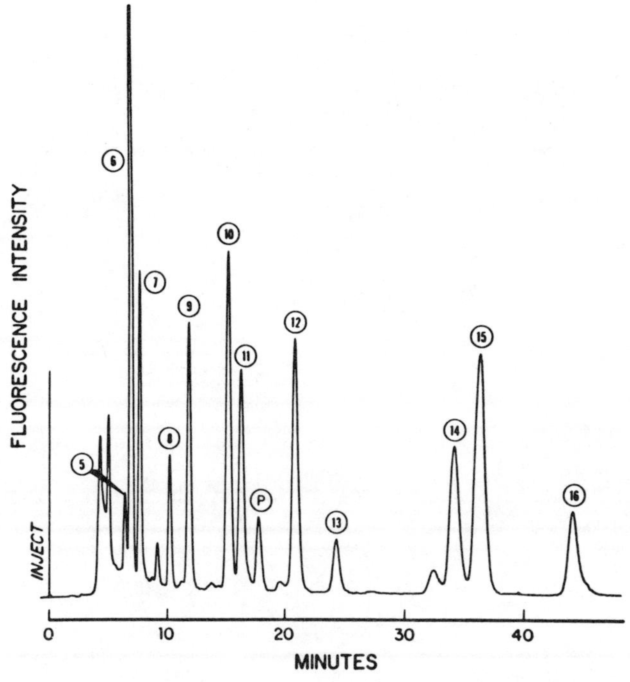

FIGURE 35. Chromatogram of 16 PAH standards using isocratic elution (80% acetonitrile in water at a flow rate of 0.5 m*l*/min) at room temperature. Peak legend: 1. Naphthalene; 2. Acenaphthene; 3. Fluorene; 4. Phenanthrene; 5. Anthracene; 6. Fluoranthene; 7. Pyrene; 8. Benz(*a*)anthracene; 9. Chrysene; 10. Benzo(*e*)pyrene; 11. Benzo(*b*)fluoranthene; P - Perylene; 12. Benzo(*K*)fluoranthene; 13. Benzo(*a*)pyrene; 14. Dibenz(*a,h*)anthracene; 15. Benzo(*ghi*)perylene; 16. Indeno(*1,2,3-cd*)pyrene. (Reprinted with permission from Ogan, K., Kate, E., and Slavin, W., *Anal. Chem.*, 51, 1315, 1979. Copyright by the American Chemical Society.)

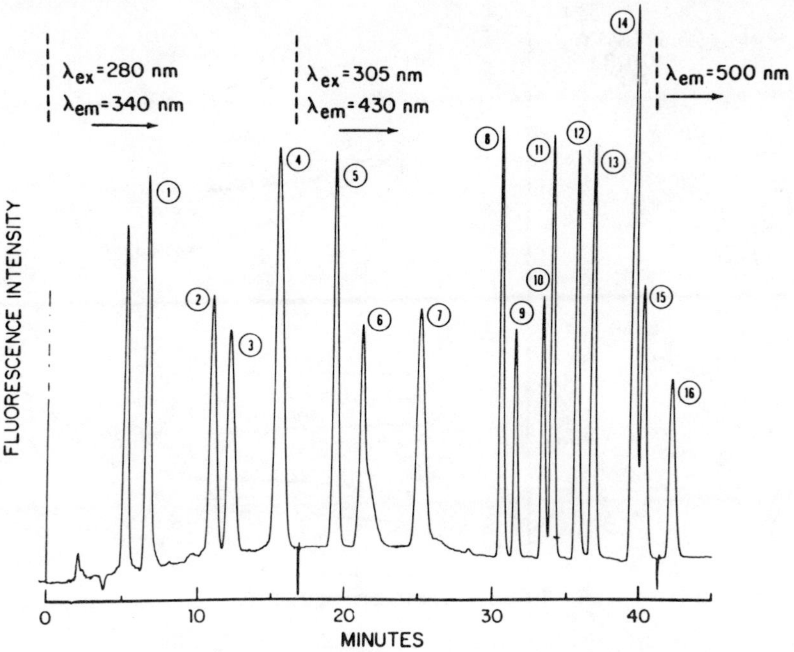

FIGURE 36. Chromatogram of 16 PAH standards using the solvent program of Figure 34, and conditions of Table 20. For peak identification, see legend in Figure 35. The tail on peak 6 is a solvent impurity. (Reprinted with permission from Ogan, K., Katz, E., and Slavin, W., *Anal. Chem.*, 51, 1315, 1979. Copyright by the American Chemical Society.)

to separate: phenanthrene/anthracene, benz(*a*)anthracene/chrysene/triphenylene and benzo(*e*)pyrene/benzo(*b*)fluoranthene/benzo(*k*)fluoranthene/benzo(*a*)pyrene. Peak shapes were sharp all the way to and including indeno[*1,2,3-cd*]pyrene, molecular weight 278, which eluted after 40 min.

Ogan et al.[140] have studied the separation characteristics of individual PAHs as a function of mobile phase composition and they found that acenaphthene and fluorene were best resolved with a reduced mobile phase strength. Mobile phase strength increases with the percentage of organic solvent in the mixture, reaching a maximum at 100% organic solvent. Dibenz(*a,h*)anthracene and benzo(*ghi*)perylene were best resolved at concentrations of acetonitrile near 100%. Relative selectivity factors, α, were calculated according to the equations:

$$\alpha = \frac{k'_1}{k'_r} \tag{8}$$

$$k' = \frac{V_1 - V_O}{V_O} \tag{9}$$

where k'_i is the capacity ratio of the particular PAH, k'_r is the value for B(a)P, V_1 is the retention volume of the solute and V_o is the dead volume. Figure 37 illustrates the change in α as the percentage of acetonitrile increases.

Retention times were strongly dependent on the length of time allowed for equilibration of the apparatus. Retention times during a single day were reproducible to 0.8 to 1.6% for PAH up to fluoranthene and 0.5 to 0.7% for PAH of higher molecular weight. Day to day precision was slightly less.

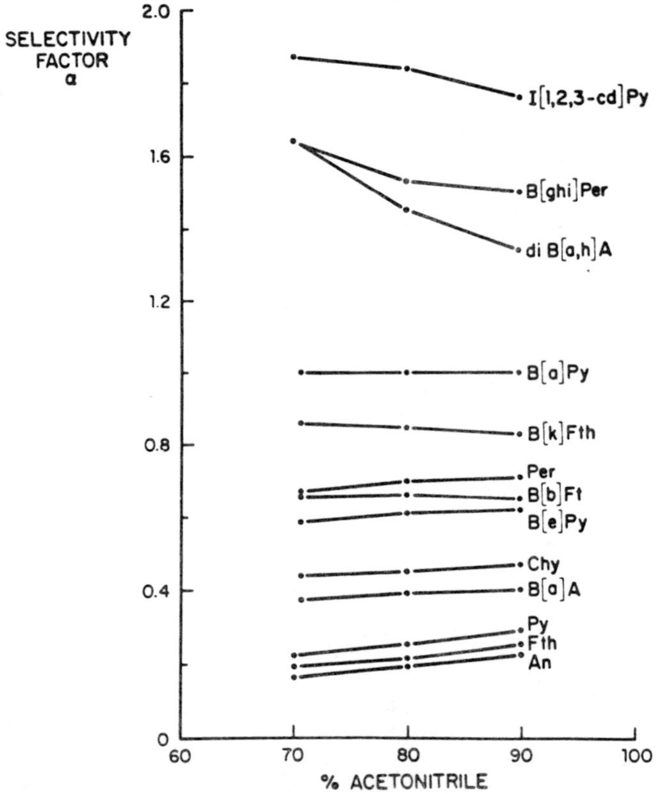

FIGURE 37. Selectivity factor, α, relative to benzo(*a*)pyrene, as a function of acetonitrile content of the mobile phase (isocratic elution). (Reprinted with permission from Ogan, K., Katz, E. and Slavin, W., *Anal. Chem.*, 51, 1315, 1979. Copyright by the American Chemical Society.)

Results for real environmental samples which were analyzed by the same procedure are shown in Figures 38 and 39. Figure 38 illustrates the analysis of an extract of flyash wash water and Figure 39 an extract of a river water sample. The quantitative detection procedures used by Ogan are described later.

Das and Thomas[142] have employed a hplc method using a 25 cm × 0.21 cm column packed with a Dupont Zorbax® ODS C_{18} stationary phase. An isocratic elution procedure was employed with a solvent mixture of 82% acetonitrile in water as the mobile phase to analyze for PAH in samples of coke oven waste water and extracts from particulate emissions. The isomers benz(*a*)anthracene/chrysene and benzo(*k*)fluoranthene/benzo(*e*)pyrene/perylene were unresolved.

Some of these same difficulties were encountered by Nielsen[119] in hplc analyses of gasoline and crankcase oil samples containing PAH. The PAH content of the samples were isolated from non-PAH fractions by means of an initial tlc separation on silica gel, details of which were given earlier. For the hplc separation, a Zorbax® ODS column, 25 cm × 0.46 cm, was employed as the analytical column with a Nucleosil® 5 C_{18} precolumn, 6 cm × 0.46 cm. After prolonged use of the analytical column, the column capacity was noted to decrease, therefore the short precolumn served a protective role. Upon incorporation of the Nucleosil® precolumn, 600 analyses resulted in no decrease in capacity. An isocratic elution mode was used with a mobile phase of methanol:water, (8:1).

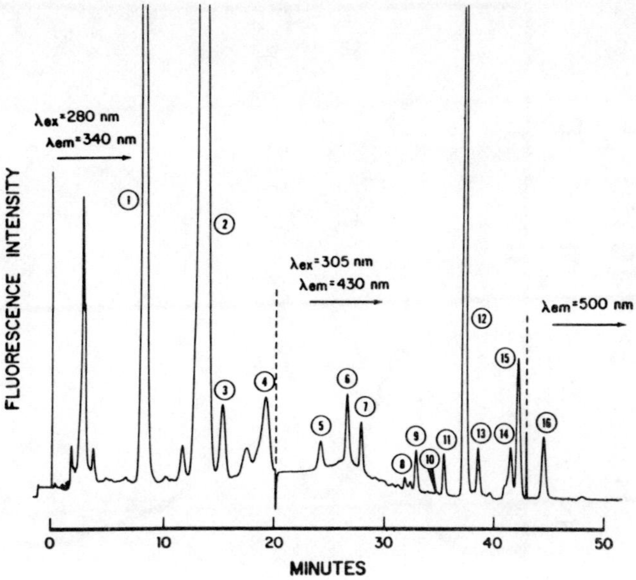

FIGURE 38. Chromatogram of the extract of flyash wash water obtained with the hplc conditions given in Table 20. For peak identification, see Figure 35. (Reprinted with permission from Ogan, K., Katz, E., and Slavin, W., *Anal. Chem.*, 51, 1315, 1979. Copyright by the American Chemical Society.)

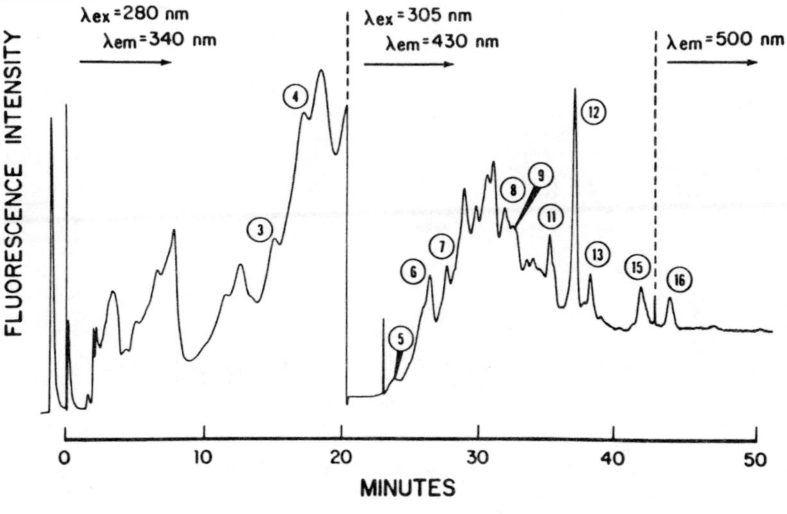

FIGURE 39. Chromatogram of the extract of a river water sample obtained with the hplc conditions in Table 20. For peak identification, see Figure 35. (Reprinted with permission from Ogan, K., Katz, E., and Slavin, W., *Anal. Chem.*, 51, 1315, 1979. Copyright by the American Chemical Society.)

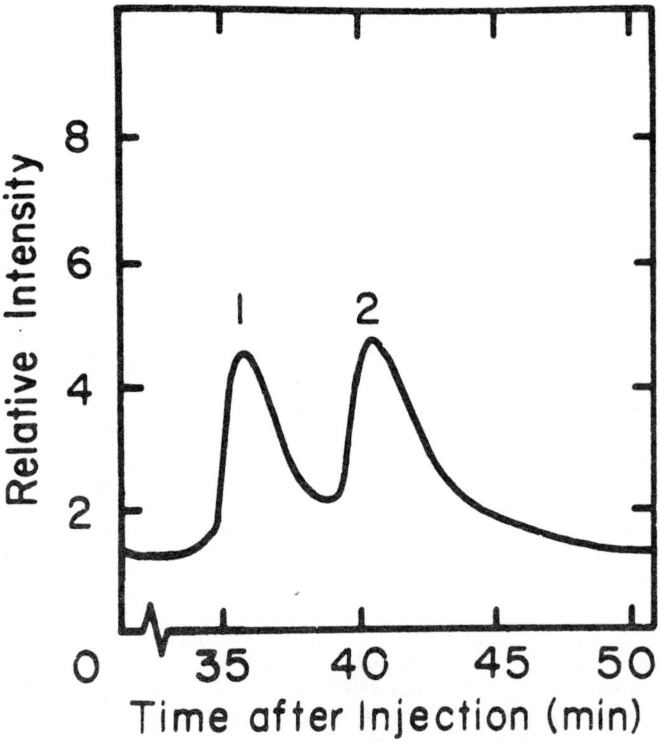

FIGURE 40. Hplc/fluorescence trace of a mixture of benzo(*e*)pyrene, (1), and benzo(*a*)pyrene, (2). Chromatographc conditions: 25 cm × 0.80 cm i.d. Zorbax® column, mobile phase, 13:7 (v/v) methanol-water, 65°C, 1400 psi, flow rate, 1.5 mℓ/min. (Reprinted with permission from Fox, M. A. and Staley, S. W., *Anal. Chem.*, 48, 992, 1976. Copyright by the American Chemical Society.)

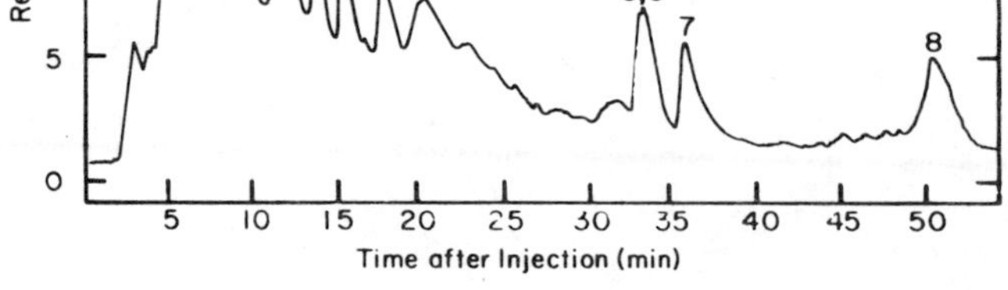

FIGURE 41. Partial hplc/fluorescence trace of a benzene extract of an atmospheric particulate matter sample collected in the Baltimore Harbor Tunnel. Chromatographic conditions: 25 cm × 0.80 cm i.d. Zorbax® ODS column, 7:3 (v/v) methanol-water, 65°C, 1600 psi, flow rate: 1.7 mℓ/min. Peak legend: 1. Fluoranthene; 2. Pyrene; 3. Benz(*a*)anthracene; 4. Chrysene; 5. Benzo(*e*)pyrene; 6. Benzo(*k*)fluoranthene; 7. Benzo(*a*)pyrene; 8. Benzo(*ghi*)perylene. (Reprinted with permission from Fox, M. A. and Staley, S. W., *Anal. Chem.*, 48, 992, 1976. Copyright by the American Chemical Society.)

Nielsen[119] has tabulated relative retention times for several PAH employing the above mentioned hplc conditions (see Table 21). He found that the benz(*a*)anthracene/chrysene/triphenylene isomers were totally unresolved, as were the 1- and 9-methyl anthracene isomers. There seemed to be a reasonable separation of the other components of the important PAH fraction: MW-252, B(e)P, B(a)P, B(k)F, and perylene, as well as an excellent separation of the two dibenzanthracene isomers.

Sorrell et al.[113] have studied the possibility of PAH contamination in drinking water from distribution pipes coated with corrosion inhibiting coal tar and petroleum based sealants. The water held in a test pipe system was subjected to liquid-liquid extraction with cyclohexane and the extract was concentrated in a Kuderna-Danish apparatus. The PAH fraction was separated on alumina prior to the hplc analysis with a Waters, μ-Bondapak® C_{18} column. Isocratic elution at a flow rate of 1 mℓ/min with a 70% acetonitrile mobile phase was used.

The cleanup procedure resulted in three fractions of PAH eluted from the alumina column. Despite this initial fractionation, benz(*a*)anthracene and chrysene as well as benzo(*b*)fluoranthene and benzo(*k*)fluoranthene were only partially resolved in the hplc step. However, the cleanup did result in the separation of perylene and B(a)P from each other as well as from the benzofluoranthenes. In addition, benzo(*ghi*)perylene, indeno(*1,2,3-cd*)pyrene, and dibenz(*a,h*)anthracene could all be identified without interference.

Fox and Staley[143] used a 25 cm × 0.8 cm stainless steel column packed with Dupont Zorbax® ODS for the analysis of PAH found in airborne particles. The PAH were eluted isocratically with a 7:3 (v/v) methanol to water mobile phase. In this investigation, benz(*a*)anthracene and chrysene could not be completely resolved although they could be distinguished by use of selective modulation techniques. Benzo(*a*)pyrene and benzo(*e*)pyrene were resolved in test samples (see Figure 40), but in environmental samples B(e)P and B(k)F were unresolved. In a chromatogram of a sample of airborne particulate matter (Figure 41), the identified PAH are numbered but other unidentified components may mask additional PAH present in smaller concentrations.

Dong et al.[144] also investigated the problem of the analysis of PAH in airborne matter using hplc conditions identical to those of Fox et al. The poor resolution of isomeric PAH that they obtained with isocratic elution strongly suggests the advisability of gradient elution for this class of separation.

Krstulovic et al.[145,146] have studied PAH in various air samples from the Rhode Island area, with a Partisil® 10-ODS column, 25 × 0.46 cm. Gradient elution was used with an initial mobile phase of 60/40 (v/v) methanol-water. Over a 34-min period, the methanol concentration was linearly increased to 100% which amounts to a rise of 1.2%/min. Fairly good separations were achieved on a model solution of 11 PAH ranging from naphthalene to coronene (Figure 42), however, the isomeric pairs B(a)P/perylene and fluoranthene/pyrene were not 100% resolved. Although there is a dependence on the column dead time and flow rate, the 1.2% rate of increase in solvent strength used by these investigators is quite low compared to the optimum value of 6.7% recommended by Dolan et al.[137] It appears that either the full advantages of gradient elution were not realized or that the μ-Bondapak® column was not extremely selective towards PAH.

Euston and Baker[92] have studied the separation of eight PAH by linear gradient and isocratic elution procedures with a LiChrosorb® RP-18 column. Although a 12 min run with a linear gradient from 60 to 95% acetonitrile in water provided a much better column efficiency than isocratic elution at 82.5% acetonitrile, they preferred the isocratic mode of separation because solvent mixing during the gradient run introduced noise. Both runs are illustrated in Figure 43. It can be seen from this figure that benzo(*e*)pyrene co-eluted with perylene and benz(*a*)anthracene with chrysene in both cases.

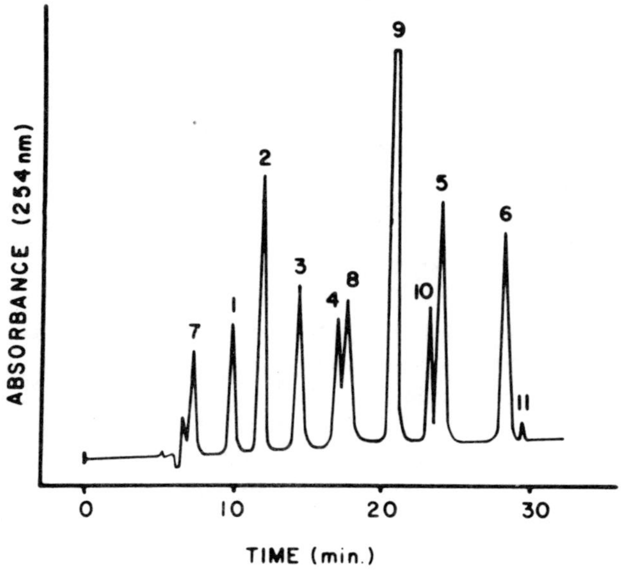

FIGURE 42. Separation of an 11 PAH standard. Chromatographic conditions: column, Partisil® 10-ODS; gradient, linear from 60/40 methanol-water (v/v) to 100% methanol in 34 min. column pressure, 630 psi, temperature, ambient. Peak Legend: 1. Benzene; 2. Naphthalene; 3. Biphenyl; 4. Phenanthrene; 5. Fluoranthene; 6. Pyrene; 7. Chrysene; 8. Perylene; 9. Benzo(*a*)pyrene; 10. Dibenz(*a,h*)anthracene; 11. Coronene. (From Krstulovic, A. M., Rosie, D. M., and Brown, P. R., *Am. Lab.*, 9, 11, 1977. With permission.)

Kasiske et al.[147] have studied separations of the six World Health Organization PAH by hplc on Nucleosil® C_{18} with a mobile phase of 100% methanol. There was poor separation of the benzofluoranthenes although benzo(*a*)pyrene was resolved as were the other three PAH. Samples from river water and waste dump seepage were analyzed.

Chmielowiec and Sawatzky[148] have performed an interesting study on the effect of temperature on hplc separations of PAH. They used a 31.4 cm × 0.46 cm column packed with Chromegabond®-C_{18} (ES Industries, Marlton, N.J.) and used acetonitrile:water (30:20 v/v) as the mobile phase. Figure 44 shows a separation of PAH carried out under isothermal conditions at 25°C and Figure 45 depicts the same hplc separation with a linear temperature gradient of 2.6°C/min up to 50°C. Compounds 13 and 14 which were unresolved under isothermal conditions were resolved when temperature programming was employed. The peak widths were sharpened and the analysis time was reduced. The thermodynamic basis for temperature effects on separations was outlined by Chmielowiec and Sawatzky and involves the consideration of both entropy and enthalpy factors. Their initial studies were the basis for their conclusion that controlled temperature variation has the potential for providing significant improvement in resolution for difficult to separate compounds, e.g., PAH. In addition, the elution of higher molecular weight PAH can be accelerated. They also suggested that instrumentation and mobile phase parameters require further study.

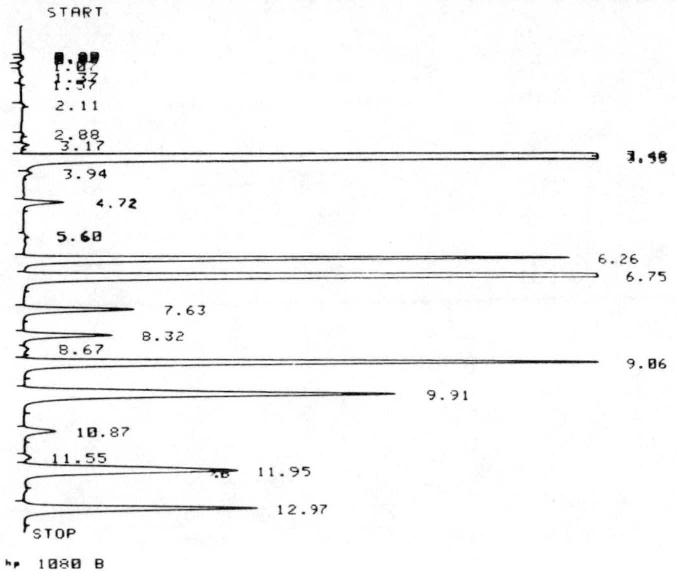

A

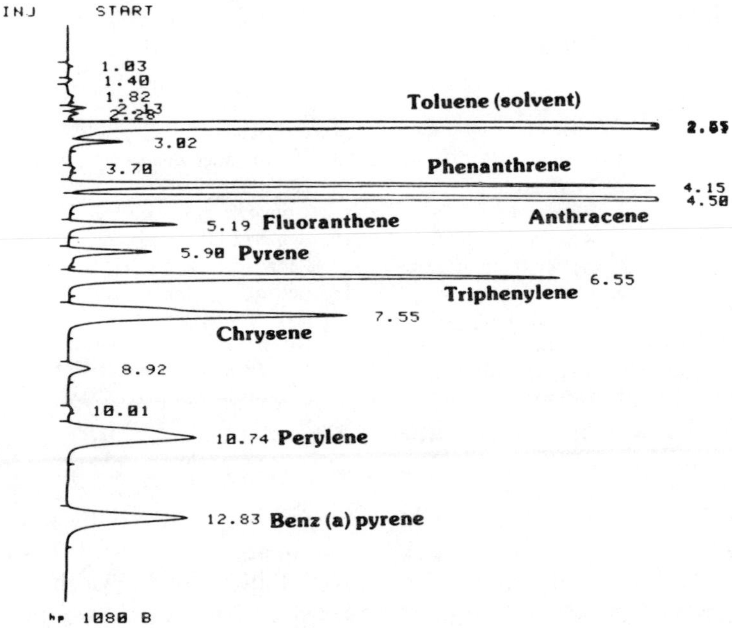

B

FIGURE 43. A = X21 B = X23
Separation of an eight PAH standard by gradient elution technique on a C_{18} reversed phase packing; gradient: 70% acetonitrile in water (v/v) to 95% acetonitrile in water in 12 min. b. Isocratic separation of the same PAH's, mobile phase, 82.5% acetonitrile in water. (From Euston, C. B. and Baker, D. R., *Am. Lab.*, 9, 91, 1979. With permission.)

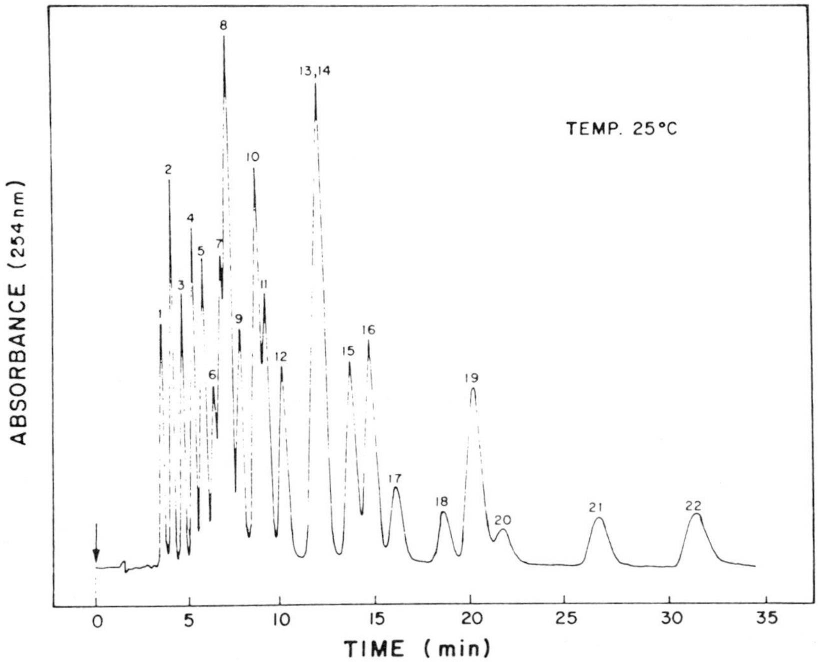

FIGURE 44. Isothermal separation of PAH on silica-C_{18} column with mobile phase of acetonitrile-water (4:1, v/v).

Peak Legend:

1. Naphalene	12. Chrysene
2. Biphenyl	13. 1,3,5-triphenylbenzene
3. Fluorene	14. 2,2′-dinaphthyl
4. Phenanthrene	15. 2-phenylanthracene
5. Anthracene	16. Benzo (*b*) fluoranthene
6. 2-phenylnaphthalene	17. Benzo (*k*) fluoranthene
7. Fluoranthene	18. Benzo (*a*) pyrene
8. 9-methylanthracene	19. 9,10-diphenylanthracene
9. Pyrene	20. Dibenz (*a,h*) anthracene
10. Triphenylene	21. Benzo (*ghi*) perylene
11. Difluorenyl	22. 1,2,3,4-dibenzopyrene

(From Chmielowiec, J., and Sawatzky, H., *J. Chromatogr. Sci.*, 17, 245, 1979. With permission.)

Lankmayr and Muller[120] have studied the separation of PAH by hplc with a LiChrosorb® RP-18 stationary phase packing and an acetonitrile-water (85:15 v/v) mobile phase. The k′ values obtained for the PAH studied are shown in Table 22. The separations at MW:228, MW:252, and MW:278 should be especially noted. Although chrysene and benz(*a*)anthracene were essentially unresolved, these two compounds were resolved from triphenylene. A partial resolution of B(b)F, perylene, B(e)P, B(k)F, and B(a)P was achieved and B(a)P was well-separated from the rest. The two dibenzanthracene isomers were completely resolved and the components of the MW:278 fraction, indeno[*1,2,3-cd*]pyrene, and benzo(*ghi*)perylene were partially distinguishable. No results were presented on the reproducibility of k′ values.

3. Alternative Reversed Phase Systems

Phthalimidopropylsilane, PPS, developed by Hunt et al.[149] is the only successful alternative reversed phase packing for hplc separations of PAH that has been em-

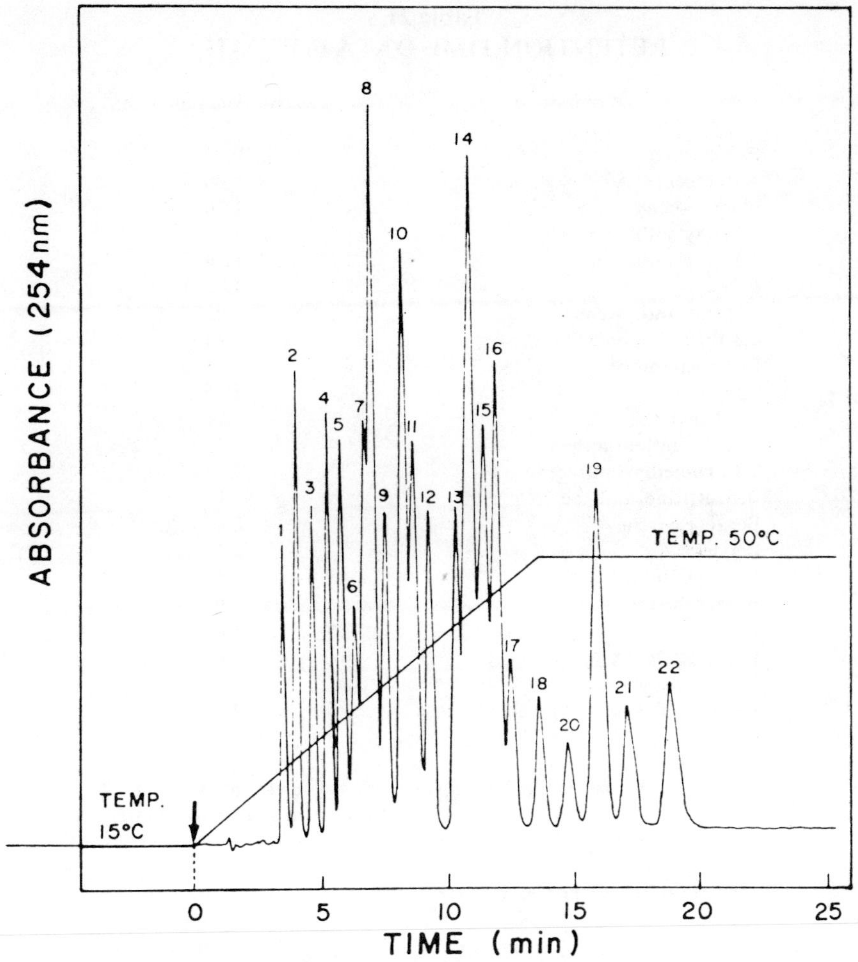

FIGURE 45. Gradient temperature separation of compounds shown in Figure 44. (From Chmielowiec, J. and Sawatzky, H., *J. Chromatogr. Sci.*, 17, 245, 1979. With permission of Preston Publications, Inc.)

ployed both in a reversed phase and normal mode for PAH separations. The structure of the starting material which is bonded to silica is shown below:

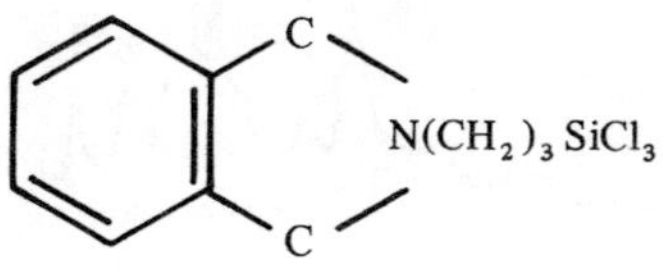

The support for this packing material is Partisil® 5, a microparticulate silica gel (5 μm), which is refluxed with 2*M* HCl for 2 hr and then sequentially rinsed with water and methanol until neutral. After overnight drying at 100°C, 7 g of the treated silica gel is reacted in a dry flask with 25 mℓ of dry xylene and 10 mℓ of 3-phthalimidopropyltrichlorosilane for 2 hr. At the end of the reflux period, 2 mℓ of methyltrichlorosilane are added and refluxing is continued for 30 min to deactivate the silica surface as completely as possible. Upon cooling, the PPS is washed thoroughly with xylene to remove unreacted silanes and is then washed with methanol.

Table 21
RETENTION TIME DATA FOR PAH

Compound	Relative retention time[a]
Phenanthrene	0.93
Anthracene	1.00
Fluoranthene	1.20
1-Methylanthracene	1.31
9-Methylanthracene	1.34
Pyrene	1.35
2-Methylanthracene	1.42
2,3-Benzofluorene	1.69
Benz(*a*)anthracene	1.75
Chrysene	1.76
Triphenylene	1.76
Benzo(*ghi*)fluoranthene	1.78
9,10-Dimethylanthracene	1.83
Benzo(*j*)fluoranthene	2.2
Benzo(*e*)pyrene	2.4
Perylene	2.5
Benzo(*k*)fluoranthene	2.6
Benzo(*a*)pyrene	2.9
Dibenz(*a,c*)anthracene	3.1
Dibenz(*a,h*)anthracene	3.6
Benzo(*ghi*)perylene	4.3
Indeno(*1,2,3-cd*)pyrene	4.6
Coronene	5.7

Note: Precolumn, Nucleosil® 5 C_{18} 6 cm × 0.46 cm; main column, Zorbax® ODS 25 cm × 0.46 cm. Solvent, methanol-water (8:1) at 1.0mℓ/min; pressure, 1700 psi, temperature, 21°.

[a] Anthracene ca. 11.5 min.

From Nielson, T., *J. Chromatogr.*, 170, 147, 1979. With permission.

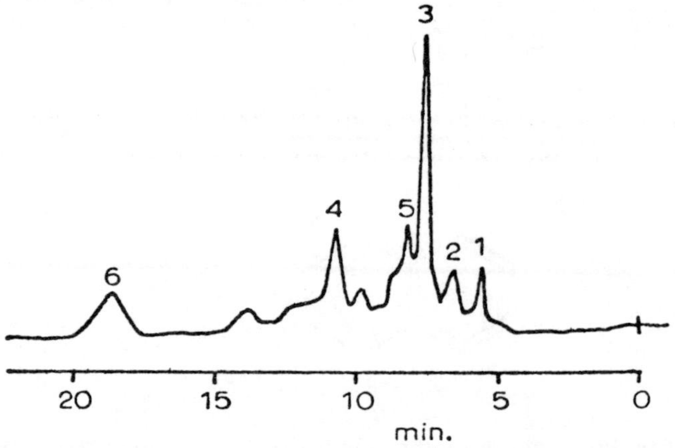

FIGURE 46. Extract of mussels chromatographed on a PPS column. 1: benz*(a)*-anthracene; 2: 12-methylbenz*(a)* anthracene; 3: benzo*(k)* fluoranthene; 4: perylene; 5: dibenz*(a,h)* anthracene; 6: benzo*(ghi)* perylene, conditions: mobile phase of 90% methanol at a flow rate of 1.5 mℓ/min; detection, fluorescence: temperature, 20°. (From Hunt, D. C., Wild, P. J., and Crosby, N. T., *J. Chromatogr.*, 130, 320, 1977. With permission.)

Table 22
k′ VALUES OF SEVERAL PAH ON LICHROSORB® RP-18

Compound	k′
Fluoranthene	2.09
Pyrene	2.71
Triphenylene	2.82
11H-Benzo(*a*)fluorene	2.88
Benz(*a*)anthracene	3.29
Chrysene	3.35
Benzo(*b*)fluoranthene	4.89
Perylene	4.99
Benzo(*e*)pyrene	5.05
Benzo(*k*)fluoranthene	5.33
Dibenz(*a,c*)anthracene	5.84
Benzo(*a*)pyrene	6.17
Dibenz(*a,h*)anthracene	7.01
3-Methylcholanthrene	9.21
Indeno(*1,2,3-cd*)pyrene	9.29
Benzo(*ghi*)perylene	9.55
Coronene	18.96

Note: LiChrosorb® RP-18 (5 μm), 150 × 3.2 mm, mobile phase: acetonitrile-water (85:15).

From Lankmayr, E. P. and Muller, K., *J. Chromatogr.*, 170, 139, 1979. With permission.

This packing was used in a 25 cm × 0.46 cm stainless steel column with a mobile phase of methanol to water (90:10). The retention time results of an hplc run on a PAH mixture are shown in Table 23. Chrysene was not one of the included compounds so that its separation from benz(*a*)anthracene cannot be evaluated. However, B(k)F, B(e)P, B(a)P, and perylene were completely separated. The chromatogram of an extract from mussels is shown in Figure 46. It would be interesting to see whether the application of a proper gradient elution procedure could further improve PAH resolutions. The detection of benzo(*ghi*)perylene in only 20 min is noteworthy.

4. Normal Phase PAH Separations

Hunt et al.[150] employed PPS as a stationary phase in a normal mode separation of PAH by hplc with a mobile phase of toluene/hexane (1:10). The six World Health Organization PAH were studied. Complete baseline separation of the six PAH: fluoranthene, benzo(*k*)fluoranthene, benzo(*b*)fluoranthene, benzo(*a*)pyrene, indeno(*1,2,3-cd*)pyrene, and benzo(*ghi*)perylene was accomplished.

Crathorne and Fielding[151] examined the separation of the six World Health Organization PAH using a Partisil®-5 packing, microparticulate silica gel. Although the benzofluoranthene isomers were only 50% resolved using a mobile phase of 0.01% acetone in hexane, resolution of all six was accomplished (see Figure 47). They noted that small changes in the mobile phase composition could dramatically affect separations, therefore care is essential in preparing the mobile phase and in activating the column. The most successful approach involved a rigorous exclusion of water from the solvents used in the mobile phase.

A number of other bonded phase packing materials have been devised for PAH separation studies. Lankmayr and Muller[120] studied silica gel supports modified with NH_2 and NO_2 groups in normal mode systems and among these, LiChrosorb® NH_2

Table 23
RELATIVE RETENTION TIMES FOR PAH ON PPS

Compound	Relative retention time
Fluoranthene	0.48
Benz(*a*)anthracene	0.55
9,10-Dimethylanthracene	0.61
12-Methylbenz(*a*)anthracene	0.64
7-Methylbenz(*a*)anthracene	0.73
Benzo(*ghi*)fluoranthene	0.75
Benzo(*b*)fluoranthene	0.82
Benzo(*k*)fluoranthene	0.82
Dibenz(*a,h*)anthracene	0.93
Benzo(*a*)pyrene	1.00
Benzo(*e*)pyrene	1.07
3-Methylcholanthrene	1.07
Perylene	1.20
Indeno(*1,2,3-cd*)pyrene	1.45
Benzo(*ghi*)perylene	1.86
Anthanthrene	1.86

Note: Conditions: mobile phase, methanol-water(90:10) at a flow rate of 1.5 mℓ/min. Retention time of benzo(*a*)pyrene was 11.5 min.

From Hunt, D. C., Wild, P. J., and Crosby, N. T., *J. Chromatogr.*, 130, 320, 1977. With permission.

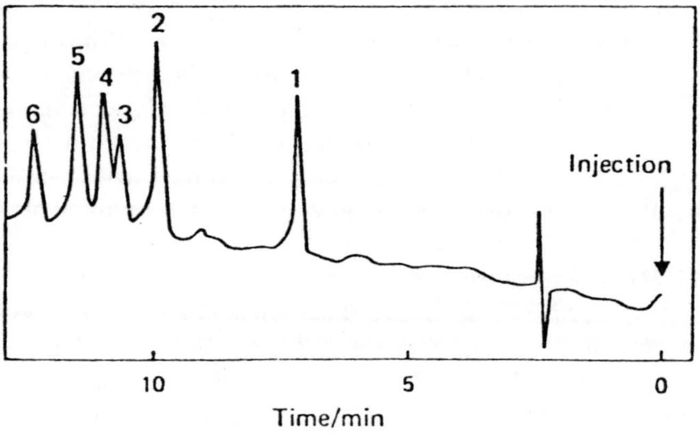

FIGURE 47. Separation of the six World Health Organization PAH. Chromatographic conditions: Column, 25 cm of Partisil®-5; mobile phase, 0.01% acetone in hexane; flow rate, 1.5 mℓ/min; ultraviolet detection, 287 nm. (From Crathorne, B. and Fielding, M., *Proc. Anal. Div. Chem. Soc.*, 15, 155, 1978. With permission.)

Peak Legend

1. Fluoranthene
2. Benzo(*a*)pyrene
3. Benzo(*k*)fluoranthene
4. Benzo(*b*)fluoranthene
5. Benzo(*ghi*)perylene
6. Indeno(*1,2,3-cd*)pyrene

Table 24
k′ VALUES OF SEVERAL PAH ON LICHROSORB® NH_2 AND NUCLEOSIL® NO_2

Compound	k′ Nucleosil® NO_2	Compound	k′ LiChrosorb® NH_2
Fluoranthene	1.45	Pyrene	1.18
Pyrene	1.47	Fluoranthene	1.48
Benz(*a*)anthracene	1.94	Triphenylene	1.86
3-Methylcholanthrene	2.09	Chrysene	2.00
Chrysene	2.16	Perylene	2.07
Triphenylene	2.33	Benz(*a*)anthracene	2.12
Benzo(*k*)fluoranthene	2.44	3-Methylcholanthrene	2.32
Benzo(*a*)pyrene	2.79	Benzo(*a*)pyrene	2.57
Benzo(*b*)fluoranthene	3.05	Benzo(*b*)fluoranthene	2.72
Benzo(*e*)pyrene	3.19	Benzo(*e*)pyrene	2.72
Perylene	3.37	Benzo(*k*)fluoranthene	2.78
Dibenz(*a,h*)anthracene	3.64	Indeno(*1,2,3-cd*)pyrene	3.38
Indeno(*1,2,3-cd*)pyrene	3.87	Benzo(*ghi*)perylene	3.52
Benzo(*ghi*)perylene	4.32	Dibenz(*a,c*)anthracene	3.55
Dibenz(*a,c*)anthracene	4.52	Dibenz(*a,h*)anthracene	3.67
Coronene	6.45	Coronene	4.60

Note: Nucleosil® 5 NO_2, 150 × 3.2 mm, mobile phase: isooctane-dichloromethane (9:1); LiChrosorb® NH_2 (10 μm), 250 × 3.2 mm, mobile phase: isooctane dried over alumina.

From Nielson, T., *J. Chromatogr.*, 170, 147, 1979; Lankmayr, E. P. and Muller, K., *J. Chromatogr.*, 170, 139, 1979. With permission.

(10 μm), and Nucleosil® 5 NO_2 are commercially available. Mobile phases were selected on the basis of good resolution and time of analysis. For LiChrosorb® NH_2, isooctane dried over alumina was chosen as the mobile phase while isooctane containing 10% dichloromethane was used for separations with Nucleosil® NO_2. Some k′ values for the PAH separated are listed in Table 24. Fluoranthene and pyrene were unresolved on Nucleosil® NO_2 although there was a partial separation of the components comprising the MW:228 and MW:252 fractions.

Lochmuller and Amoss[152] developed a bonded charge transfer type of stationary phase packing for the separation of alkyl-substituted PAH. Vespalec[153] has investigated the selectivity of 1,2,3-tris(2-cyanoethoxy)propane coated as a liquid on silica gel for the separation of PAH. The separations of isomeric PAH were poor in both cases.

Popl et al.[154-156] have studied the effect of the molecular structures of various PAH on their adsorptivity on silica gel and alumina using a normal phase system with dry n-pentane as the mobile phase. Adsorption energies of PAH were calculated from measured retention volumes. Retention indices were calculated in a manner analogous to the Kovats' index system for gas chromatography. The studies with alumina and silica gel focused primarily on the relationship of retention indices to factors such as type of ring system, alkyl branching, and carbon number. Popl et al.[156] have extended this form of systematic study towards the chromatography of aromatic hydrocarbons on macroporous polystyrene gel.

Wise et al.[157] at the National Bureau of Standards have carried out studies which will enable predictions to be made regarding separations on the C_{18} phases, which are the most commonly employed for PAH separations. They have devised a two-stage analytical scheme for analyses of PAH in complex environmental samples such as crude oil fractions. In the first step, a polar chemically bonded stationary phase, μ-

Table 25
RETENTION INDICES FOR AROMATIC HYDROCARBONS ON TWO STAGE μ-BONDAPAK® NH_2 — μ-BONDAPAK® C_{18} SYSTEM

Compound	Logarithm of the retention index (log I)	
	μ-Bondapak® NH_2[a]	μ-Bondapak® C_{18}[b]
Two-ring aromatics		
Naphthalene	2.00	2.00
2-Methylnaphthalene	1.96	2.59
2,3-Dimethylnaphthalene	2.08	3.47
2,6-Dimethylnaphthalene	1.94	3.15
1,5-Dimethylnaphthalene	1.93	3.00
2,3,6-Trimethylnaphthalene	2.00	3.54
1,4,6,7-Tetramethylnaphthalene	2.05	3.86
2-Ethylnaphthalene	1.92	3.14
Biphenyl	2.25	2.57
3-Ethylbiphenyl	2.20	3.46
Acenaphthene	2.00	2.73
Three-ring aromatics		
Fluorene	2.61	2.78
Dibenzothiophene	2.75	2.98
Anthracene	2.95	3.02
2-Methylanthracene	2.92	3.63
9,10-Dimethylanthracene	2.95	3.82
Phenanthrene	3.00	3.00
1-Methylphenanthrene	2.98	3.43
p-Terphenyl	3.40	3.94
Four-ring aromatics		
Fluoranthene	3.39	3.42
Benzo(*a*)fluorene	3.46	3.74
Benzo(*b*)fluorene	3.53	3.78
Pyrene	3.68	3.51
Naphthacene	3.93	—
Benz(*a*)anthracene	4.00	4.00
Triphenylene	4.00	3.82
Chrysene	4.03	3.94
Five-ring and larger aromatics		
Benzo(*a*)pyrene	4.30	4.57
Perylene	4.47	4.43
Benzo(*ghi*)perylene	4.61	> 5
Indeno(*1,2,3-cd*)pyrene	4.72	> 5
Dibenz(*a,c*)anthracene	4.93	4.84
Dibenz(*a,h*)anthracene	4.93	> 5
Picene	> 5	4.93
Benzo(*b*)chrysene	5.00	5.00

[a] Hexane mobile phase (pentane produces identical chromatographic results but flow reproducibility is poorer).
[b] Acetonitrile-water mobile phase.

Bondapak® NH_2, (an amino-propylsilane bonded chemically to microporous silica, 10 μm) was used. Polar compounds were separated on this material. When a nonpolar mobile phase such as hexane was used with this column, a hydrocarbon class fractionation was achieved. Saturated hydrocarbons eluted before the aromatic hydrocarbons and a linear relationship existed between the elution volume and the number of aromatic rings. With this separation procedure, PAH fractions, based on the number of rings, were obtained. The logarithms of the retention indices have been calculated in the same manner as Popl et al.[154,155] and are presented in Table 25.

The retention of a particular PAH on μ-Bondapak® NH_2 was affected only minimally by alkyl side chains. This is in contrast to separations on silica or alumina where the retention increases upon addition of alkyl groups. The reason for this behavior is explained in terms of the retention mechanism. In normal phase hplc, retention times are governed solely by the interaction between the π electrons of the PAH and the π electron system of the polar amino group, therefore little increase in retention is expected.

The second stage of Wise's work involved the separation on μ-Bondapak® C_{18}. A linear gradient mobile phase program was employed ranging from 50 to 100% acetonitrile in water. The gradient was spanned in 30 min. Wise's group reported an excellent resolution of a number of alkyl homologues, however, the isomeric PAH such as benz(*a*)anthracene and chrysene, and phenanthrene and anthracene were not well-resolved. Retention indices for the analysis on the C_{18} column are listed in Table 25.

C. Methods of Detection

The two most common methods of detection are based on the strong UV absorption by PAH and their characteristically high fluorescence intensities. The fluorescence method is more sensitive and selective towards PAH. Advantages of the fluorescence method are as follows: (1) the fluorescence technique allows the selection of two characteristic wavelengths, one for excitation and one for emission, (2) for strongly fluorescing substances such as PAH the fluorimetric method is more sensitive to lower concentration levels than UV methods, and (3) since not all compounds fluoresce, all potentially UV absorbing, nonfluorescing interferences are not detected. The sensitivity advantage of fluorescence detection over UV detection is illustrated in Figure 48.

The simplest approach for detection of the PAH eluants from hplc is to use an UV detector set at 254 nm. If the column is capable of completely resolving the PAH the detector will give a signal for every compound. However, since no column is able to afford complete resolution, interference effects will always exist.

The efficiency of UV detection can be improved by selectively monitoring more than one wavelength, however, a prior knowledge of the components of the sample is necessary in order to preselect the optimum UV absorption wavelength of each PAH. Krstulovic et al.[145] used six different wavelengths to analyze a mixture of 10 PAH. Figure 49 illustrates the relative sensitivity of various PAH at different wavelengths.

The main advantage of UV detectors is that they are simple and easy to operate. In practice, their usefulness for PAH detection of hplc eluants is generally confined to a subordinate role. Ultraviolet detectors are often used in series with fluorescence detectors as a means of providing a general response before the fluorescence detector provides a more selective response.

There are a number of commercially available fluorescence detectors, however, all of them can be classified into two major types. The first which is the simplest and least expensive is a modified filter fluorimeter which allows excitation and emission at set wavelengths. A set of interchangeable filters allows a reasonable choice of operating wavelengths. A more sophisticated detector can be built around a spectrofluorimeter with its variable wavelength monochromator. Any emission wavelength is then available, best detected with a high through-put, cut-off filter.

A typical optical arrangement for fluorescence detection devised by Slavin et al.[158] is shown in Figure 50. The optical system included a xenon capillary flash lamp light source, an interchangeable filter for choosing excitation wavelengths, and a continuously variable interference filter for choosing an emission wavelength. The spectrum bandpass could be limited to roughly 14 nm. The xenon lamp provided high light intensities from 260 to 656 nm. Emission wavelengths of 390 to 750 nm could be selected with the variable interference filter. A reference photomultiplier served as a

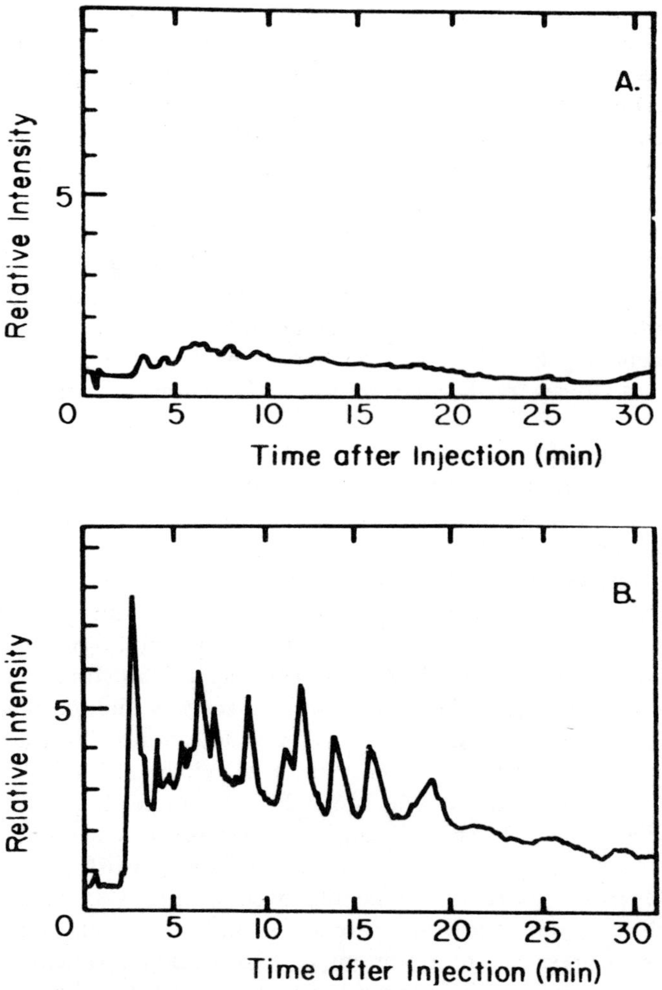

FIGURE 48. Partial hplc/UV absorption (A) and fluorescence (B) traces of a benzene extract of atmospheric particulate matter collected in the Baltimore Harbor Tunnel. These traces were determined simultaneously. Chromatographic conditions: 25 cm × 0.8 cm i.d. Zorbax® ODS column, 13:7 (/v) methanol-water, 65°C, 1400 psi, flow rate, 1.5 mℓ/min. (Reprinted with permission from Fox, M. A. and Staley, S. W., *Anal. Chem.*, 48, 992, 1976. Copyright by the American Chemical Society.)

monitor to correct for any intensity variations in the wandering of the xenon lamp. A rectangular quartz flow cell was used since this geometrical shape was claimed to produce less stray light than a circular flow cell.[158]

Two general methods of determining the fluorescence intensities of the individual PAH are available. The most common method is the dynamic method in which the column effluent passes through a flow cell, typically 5 to 50 $\mu\ell$ in volume. However, a static method is also used; in this method the fluorescence spectra are obtained by stopping the chromatographic flow and measuring the fluorescence signal.

When all of the operating parameters have been established, a calibration curve is set up to determine the working range of concentration. A standard should be available for each PAH present. The need for calibration is especially important in fluorescence

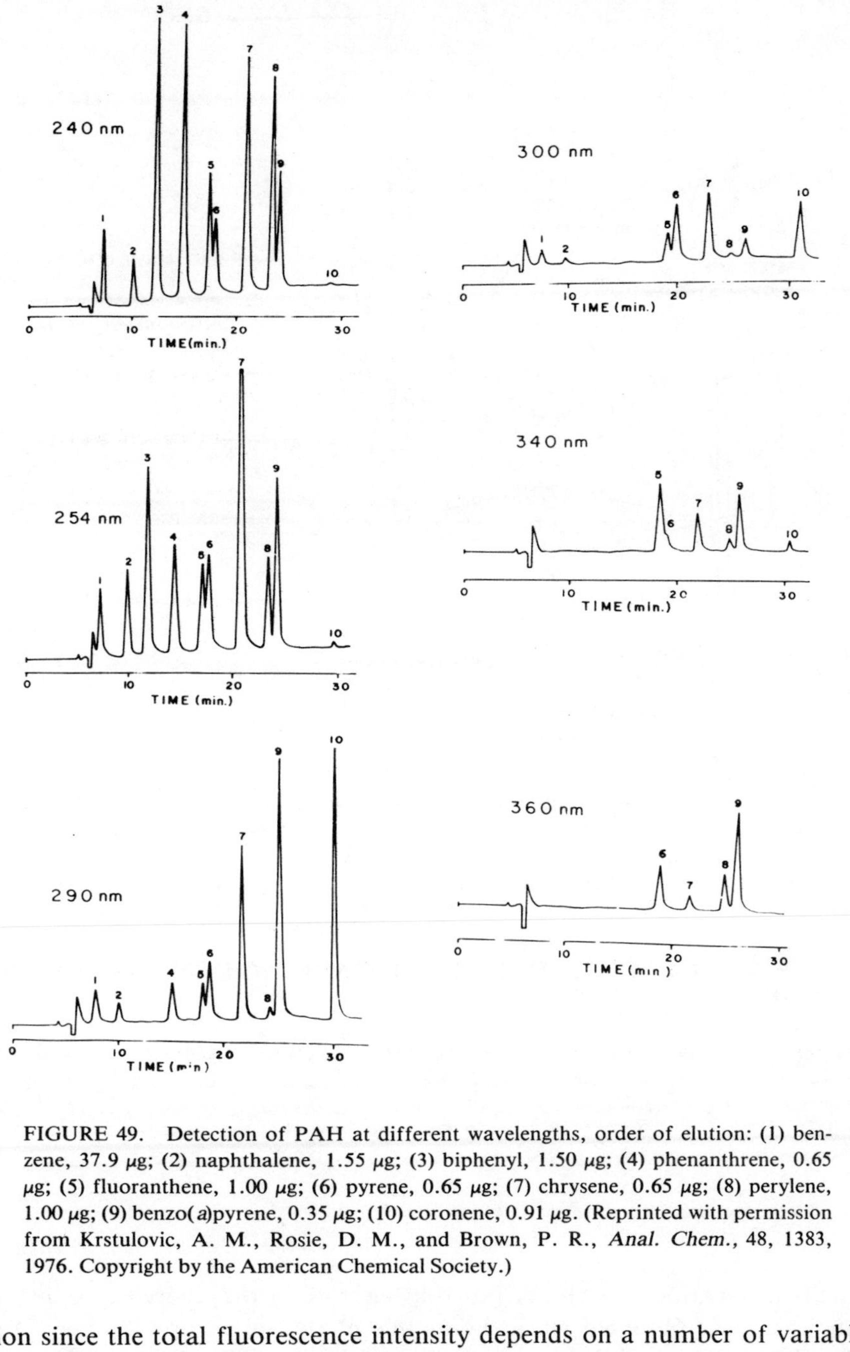

FIGURE 49. Detection of PAH at different wavelengths, order of elution: (1) benzene, 37.9 μg; (2) naphthalene, 1.55 μg; (3) biphenyl, 1.50 μg; (4) phenanthrene, 0.65 μg; (5) fluoranthene, 1.00 μg; (6) pyrene, 0.65 μg; (7) chrysene, 0.65 μg; (8) perylene, 1.00 μg; (9) benzo(*a*)pyrene, 0.35 μg; (10) coronene, 0.91 μg. (Reprinted with permission from Krstulovic, A. M., Rosie, D. M., and Brown, P. R., *Anal. Chem.*, 48, 1383, 1976. Copyright by the American Chemical Society.)

detection since the total fluorescence intensity depends on a number of variables: (1) the filter quality, (2) the purity of the mobile phase, (3) the quantum yield or fluorescence efficiency, (4) the solute fluorescence lifetime, and (5) potential quenching effects by other dissolved species such as oxygen. Self-absorption or inner quenching of the fluorescing species can affect the detector response at high concentrations but this is generally not a problem at the low concentrations of PAHs found in water, even after preconcentration steps.

The problem of fluorescence quenching by dissolved oxygen has not been widely studied. Fox and Staley[143] have noted that dissolved oxygen in solution can quench

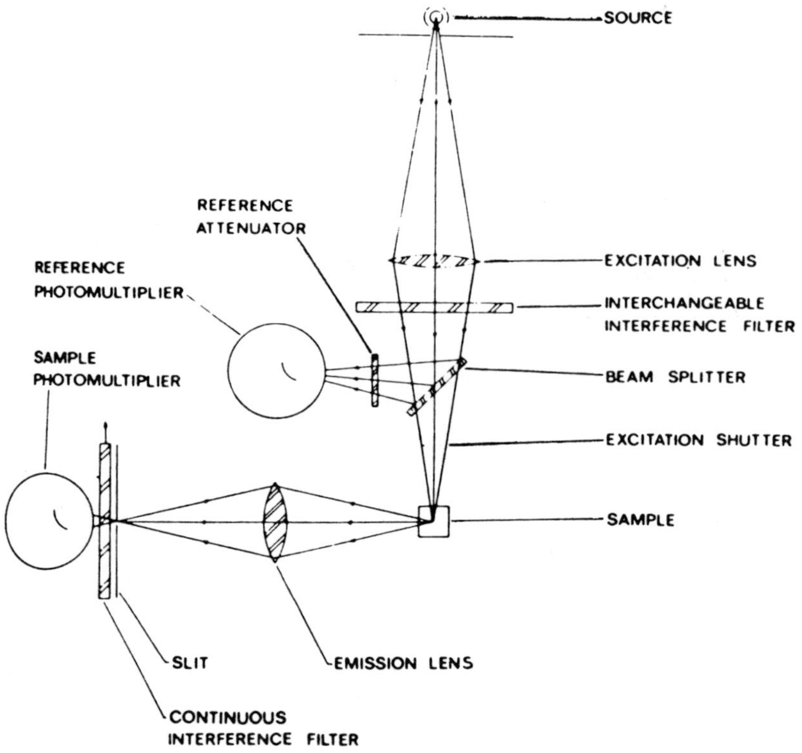

FIGURE 50. Typical optical arrangement of an hplc/fluorescence detector. (From Slavin, W., William, A. T. R., and Adams, R. F., *J. Chromatogr.*, 134, 121, 1977. With permission.)

Table 26
COMPARISON OF METHODS FOR DETERMINING DETECTION LIMITS

Criterion for determining limit	Naphthalene	Phenanthrene	Pyrene	Chrysene	Benzo(*a*)pyrene
2X Range	0.16	0.031	0.10	0.056	0.039
3X Std. Dev.	0.12	0.024	0.07	0.042	0.029
3X RMS Area	0.19	0.021	0.085	0.042	0.025

Note: UV absorbance detector at 254 nm. Detection limit in ng injected.

the excited singlets of the PAH. A fourfold increase in the fluorescence intensity of benzo(*a*)pyrene was observed when the mobile phase was completely deoxygenated. The difficulty in reproducing the deoxygenation of the mobile phase may be the explanation of the small number of published studies in this area.

It is necessary to compare the detection limits and sensitivities of PAH on the basis of available hplc detectors and the inherent problems in UV and fluorescence detection. Christensen and May[159] of the National Bureau of Standards have compared detection limits of PAH for fixed wavelength UV, variable wavelength, filter fluorimeter, and spectrofluorimetric detectors. In order to present meaningful results, they address the problem of defining sensitivity and detection limit. It was noted that sensitivity is most often expressed in terms of detection limits which are then defined in different manners in the literature. One convention is to define the detection limit as "the smallest possible amount of material which gives rise to an observable peak". Detection limits based on this interpretation depend on the type of chromatography employed as well as on the actual instrument and detector. To determine the sensitivity for a specific detector, one must know the actual concentration of material present in the detector sample cell which gives rise to an observable peak.

Ambiguities of the definition may result because an "observable peak" can be defined in a number of ways. The usual criterion is that a peak is observed if the signal to noise ratio is greater than or equal to two.[159] Christensen and May noted that even the definition of noise is subject to variation. They suggest that in the case of random noise, twice the noise range might be as many as six times the standard deviation about the mean. As a result, the probability that a peak is real, is greater than 99.9%, a criterion the authors consider too stringent. The noise level can also be characterized by the standard deviation of the signal about the mean or baseline. A real peak with a probability of 99.7% is then considered to be any peak greater than three times the standard deviation. Another definition of noise level is commonly used with electronic integrators, where the noise level is defined as the root mean square average of the peak areas produced by the noise excursions. Any peak three times greater in area than this average can be considered a true peak. The probability is also 99.7% in this case. For each of these definitions, the limits of detection for PAH using a 254 nm UV detector are compared in Table 26.

Christensen and May[159] have applied their definitions of detection limits to studies of the efficiencies of several hplc detectors for the determination of PAH. Limits of detection for a fixed-wavelength (254 nm) UV detector, three variable-wavelength UV detectors, three filter fluorimeters, and one spectrofluorimeter are presented in Tables 27 and 28. The detection limits were calculated in terms of the amount of PAH injected into the chromatograph and alternatively, the concentration of PAH in the detector cell. Several commercially available detectors denoted by A, B, C, etc., were used by NBS in these studies. Picogram quantities of PAH were observed with all of the detectors. Christensen and May concluded that the optimum detector system consists of a 254 nm fixed UV-wavelength detector in series with a spectrofluorimeter. The UV detector gives a general response to all PAH and the spectrofluorimeter can be used for the quantitative determination of individual PAH.

Ogan et al.[140] used a Perkin-Elmer Model 650-10LC fluorescence detector for the determination of PAH separated by the hplc method discussed earlier. The excitation and emission spectra of authentic samples of the 16 PAH studied were recorded. The important emission and excitation peaks of the 16 PAH are listed in Table 29. The overlapping of the spectra of some PAH necessitates the selection of compromise wavelengths for several PAH. Three groups of PAH each with distinct excitation and emission wavelengths were set up (see Table 30). These groups were distinguishable on the basis of retention times as well as on the basis of the change in selected excitation and emission wavelengths between phenanthrene and anthracene. An enhanced fluorescence intensity of indeno(*1,2,3-cd*)pyrene was realized by shifting the emission wavelength to 500 nm. The addition of standard PAH to sample extracts served as a means of positive identification of specific peaks. Since the assumption was made that each peak resulted from a single compound, a fluorescent impurity would cause an error in the quantitative determination.

The detection limit for each PAH was calculated for the solute peaks with a minimum observable peak height defined as equal to a signal to noise ratio of two. These results are presented in Table 31 and it should be noted that subpicogram amounts of acenaphthene and benzo(*k*)fluoranthene were detectable.

Das and Thomas[142] have also reported the determination of picogram and subpico-

Table 27
DETECTION LIMITS IN TERMS OF NANOGRAMS INJECTED INTO CHROMATOGRAPH

Detector	Naphthalene	Phenanthrene	Pyrene	Chrysene	Benzo(*a*)pyrene
Fixed-wavelength at 254 nm	0.16	0.025	0.085	0.046	0.021
Variable-wavelength detectors:					
A at 254 nm	4.7	0.33	1.29	0.42	0.28
B at 254 nm	0.8	0.06	0.20	0.10	0.05
C at 254 nm	4.0	0.26	0.9	0.38	0.5
A at optimum	0.14	0.20	0.26	0.19	—
C at optimum	0.21	—	0.3	0.2	—
D at optimum	0.32	—	0.07	0.12	—
Fluorimeters:					
E	—	—	—	—	0.02
F	—	—	—	—	0.007
G	—	—	—	—	0.05
Spectrofluorimeter	6.8	0.5	0.11	0.16	0.012

Table 28
DETECTION LIMITS IN TERMS OF CONCENTRATION PRESENT IN THE CELL, ng/mℓ

Detector	Naphthalene	Phenanthrene	Pyrene	Chrysene	Benzo(*a*)pyrene
Fixed-wavelength at 254 nm	0.34	0.035	0.14	0.054	0.022
Variable-wavelength detectors:					
A at 254 nm	7.8	0.55	2.2	0.47	0.19
B at 254 nm	140	10	40	10	4
C at 254 nm	6.6	0.43	1.5	0.45	0.35
A at optimum	0.23	0.33	0.43	0.21	—
C at optimum	0.35	—	0.50	0.23	—
D at optimum	0.53	—	0.11	0.13	—
Fluorimeters:					
E	—	—	—	—	0.014
F	—	—	—	—	0.005
G	—	—	—	—	0.003
Spectrofluorimeter	0.23	0.8	0.18	0.18	0.008

gram amounts of PAH by the fluorescence detection of hplc eluates. A Model FS 970 Spectrofluoro Monitor® (Schoeffel Instr. Corp.) was equipped with a low volume flow cell (5 $\mu\ell$) and a continuously variable source of monochromatic excitation energy was provided by a highly stabilized deuterium lamp (190 to 400 nm). Six emission cut-off filters, 370, 389, 418, 470, 550, and 580 nm were used to select the desired emission wavelengths.

After preliminary experiments to determine the optimum value of the excitation and emission wavelengths for each PAH, detection limits were estimated from a chromatogram by considering an observable peak to have a minimum signal to noise ratio of two. The limits for nine common PAH are listed in Table 32. The limits for chrysene and perylene were determined on individual samples at the noted experimental parameters.

Table 29
PAH PEAK WAVELENGTHS IN EXCITATION AND EMISSION SPECTRA

Compound	λ_{ex}[a]	Emission peaks	Excitation peaks	λ_{em}[a]
Naphthalene	250	330	280	340
Acenaphthene	275	328	300	340
Fluorene	260	315	265	310
Phenanthrene	295	355, 365	250, 290, 325	360
Anthracene	305	385, 405, 425, 505	250, 350, 370	405
Fluoranthene	305	340, 485	245, 290, 330, 350	480
Pyrene	295	375, 390, 410	245, 290, 325	390
Benz(*a*)anthracene	295	390, 410	285, 340	410
Chrysene	295	365, 385, 405	270, 315	385
Benzo(*e*)pyrene	305	395	290, 330	395
Benzo(*b*)fluoranthene	300	500	280, 330	500
Benzo(*k*)fluoranthene	295	415, 435	250, 305	410
Benzo(*a*)pyrene	305	410, 430, 460	270, 295, 370, 390	410
Dibenz(*a,h*)anthracene	300	395, 430, 440	292, 350	400
Benzo(*ghi*)perylene	305	340, 420	290, 370, 385	420
Indeno(*1,2,3-cd*)pyrene	305	480, 500	250, 300, 360	500

[a] Wavelengths in nm.

Calibration curves were constructed to test the linearity of the detector response. Concentrations from 1 to 100 pg of each PAH in the mixture gave a linear fluorescence response with correlation coefficients between 0.999 and 1.000.

The determination of PAH using fluorescence detection presents a number of problems since qualitative identification cannot be made without the use of standard samples. The peaks in real samples were identified by the method of standard addition. The identifications were made by use of a differential fluorescence technique in which samples and standards were analyzed under three optimal fluorescence conditions. The final characterization of peaks was accomplished by determining the peak height ratios at two distinct excitation and emission wavelengths (λ_{ex} 250/λ_{em}>370) and comparing with ratios obtained on standards.

Das and Thomas[142] used the principle of selective excitation to determine individual concentrations of chrysene and benz(*a*)anthracene, B(a)A. The use of a column packing capable of resolving the two isomers would be the easiest method for their determination but a fluorescence method can be attempted when resolution is impossible. Since chrysene exhibits a very weak UV absorption at a wavelength of 280 nm, a very weak fluorescence is emitted with this excitation wavelength. Therefore, benz(*a*)anthracene can be selectively determined in the presence of chrysene up to a

Table 30
THREE GROUPS OF PAH AND THE APPROPRIATE COMPROMISE EXCITATION AND EMISSION WAVELENGTHS

1. λ_{ex} = 280 nm, λ_{em} = 340 nm
   Naphthalene
   Acenaphthene
   Fluorene
   Phenanthrene
2. λ_{ex} = 305 nm, λ_{em} = 430 nm
   Anthracene
   Fluoranthene
   Pyrene
   Benz(*a*)anthracene
   Chrysene
   Benzo(*e*)pyrene
   Benzo(*b*)fluoranthene
   Benzo(*k*)fluoranthene
   Benzo(*a*)pyrene
   Dibenz(*a,h*)anthracene
   Benzo(*ghi*)perylene
3. λ_{ex} = 305 nm, λ_{em} = 500 nm
   Indeno(*1,2,3-cd*)pyrene

Reprinted with permission from Ogan, K., Katz, E., and Slavin, W., *Anal. Chem.*, 51, 1315, 1979. Copyright by the American Chemical Society.

Table 31
DETECTION LIMIT FOR EACH PAH AS CALCULATED FROM THE SIGNAL-TO-NOISE RATIO OF THE INDIVIDUAL PEAK IN CHROMATOGRAMS

Compound	Detection limit, pg (compromise wavelengths)
Naphthalene	30
Acenaphthene	0.5
Fluorene	12
Phenanthrene	130
Anthracene	120
Fluoranthene	40
Pyrene	75
Benz(*a*)anthracene	35
Chrysene	70
Benzo(*e*)pyrene	45
Benzo(*b*)fluoranthene	3
Benzo(*k*)fluoranthene	0.3
Benzo(*a*)pyrene	2.5
Dibenz(*a,h*)anthracene	4
Benzo(*ghi*)perylene	9
Indeno(*1,2,3-cd*)pyrene	8

Reprinted with permission from Ogan, K., Katz, E., and Slavin, W., *Anal. Chem.*, 51, 1315, 1979. Copyright by the American Chemical Society.

Table 32
DETECTION LIMITS OF STANDARD PAH

Compound	λ_{ex}(nm)	λ_{em}(nm)	Amount, pg
Fluoranthene	280	>389	0.5
Benz(*a*)anthracene	280	>389	0.6
Benzo(*k*)fluoranthene	280	>389	0.4
Benzo(*e*)pyrene	280	>389	5.1
Benzo(*a*)pyrene	280	>389	1.1
Dibenz(*a,h*)anthracene	280	>389	2.3
Benzo(*ghi*)perylene	280	>389	3.0
Chrysene	250	>370	2.3
Perylene	250	>370	0.6

B(a)A/chrysene ratio of 0.025 (5 pg B(a)A to 200 pg chrysene) at an excitation wavelength of 280 nm with an emission cutoff wavelength of 389 nm (λ_{em}>389 nm). Below this ratio of 0.025, the chrysene contribution becomes significant. This ratio can be reduced to 0.01 if the emission of B(a)A is monitored above 418 nm, but this procedure results in a loss of sensitivity.

It is just as important to be able to measure chrysene alone or in a mixture with B(a)A. Chrysene and B(a)A can both be excited at 250 nm. If the emission is monitored with a cutoff filter so that λ_{em}>370 nm, chrysene can be determined in the presence of B(a)A by the use of the following equation: $a = b - c/d$ where a = peak height due to chrysene at λ_{ex} = 250 nm and λ_{em}>370 nm, b = peak height due to both B(a)A and chrysene at λ_{ex} = 250 nm and λ_{em}>370 nm, c = peak height of B(a)A at λ_{ex} = 280 nm and λ_{em}>389 nm and d = 3.91, which is the ratio of peak heights of B(a)A at known concentrations at λ_{ex} = 280 nm λ_{em}>389 nm and λ_{ex} = 250 nm, λ_{em}>370 nm. Das and Thomas tested the validity of this equation by a series of experiments in which chrysene/B(a)A ratio varied from 0.36 to 4.36. The constant 3.91 is invariant over changes in hplc parameters as long as the fluorescence measurements at 250 and 280 nm are made under identical hplc conditions. Das and Thomas have also applied this mixture analysis to coeluting peaks at MW:252; perylene and benzo(*a*)pyrene.

Eisenbeiss et al.[87] have also studied the fluorescence properties of PAH isolated by hplc. Optimum excitation and emission wavelengths for the analysis of each component were determined from individual spectra. These data were used to establish limits of detection which were on the nanogram to subnanogram level. For six World Health Organization PAH studied, an excitation wavelength of 367 nm and an emission wavelength of 470 nm was considered a good compromise for the entire group.

Fox and Staley[143] have also used fluorescence detection for quantitation of PAH separated by hplc. In addition to conventional fluorescence detection of the separated PAH, Fox and Staley used first and second derivative fluorescence spectroscopic techniques to improve peak resolution. For the case of difficult to resolve chromatographic peaks, such as the chrysene/benz(*a*)anthracene fraction, they employed selective modulation fluorescence techniques. The fluorescence emission spectrum of the hplc fraction obtained from an airborne particulate sample, which corresponds in terms of retention time to the expected chrysene fraction, is shown in Figure 51. The standard emission spectra of pure chrysene and benz(*a*)anthracene are shown in Figures 52a and 52b. Modulation of the excitation monochromator at the maximum absorption peak

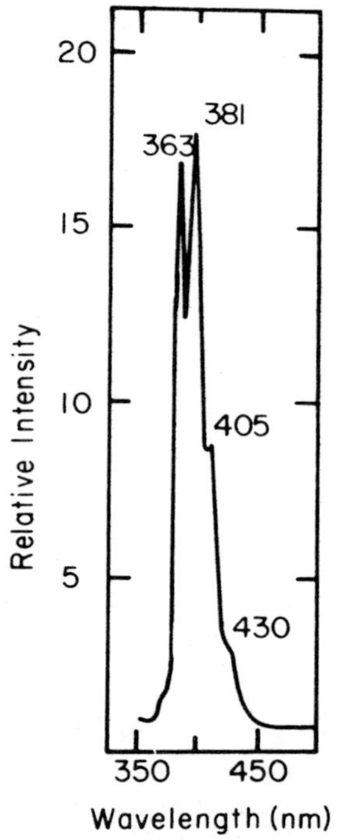

FIGURE 51. Application of selective modulation to enhance the resolution of chrysene and benz(*a*)anthracene by hplc. Fluorescence emission spectrum of a fraction of a Baltimore Harbor Tunnel sample consisting of a mixture of chrysene and benz(*a*)anthracene, λ_{exc} = 298 nm. (Reprinted with permission from Fox, M. A. and Staley, S. W., *Anal. Chem.*, 50, 967, 1978. Copyright by the American Chemical Society.)

of chrysene will effectively nullify the contribution of chrysene to the peak in Figure 51. The results of such a selective modulation technique applied to the fraction shown in Figure 51 are shown in Figure 53. If the peak in Figure 53 had consisted of only pure chrysene the result would be identical to curve 2 of Figure 53 which was obtained in a test run on pure chrysene. However, upon application of the technique, curve 1 of Figure 53 resulted. Curve 1 was identified as being identical to the spectrum of pure benz(*a*)anthracene except for an impurity at 434 nm. This impurity was traced to the solvent. In the same fashion, other coeluting PAH, e.g., B(e)P and B(k)F, were determined.

D. Future Outlook

As far as detection systems for hplc are concerned, the reported success in measuring picogram amounts of PAH is very impressive. The present success in detecting these levels has been accomplished with conventional deuterium, tungsten-halogen, and ex-

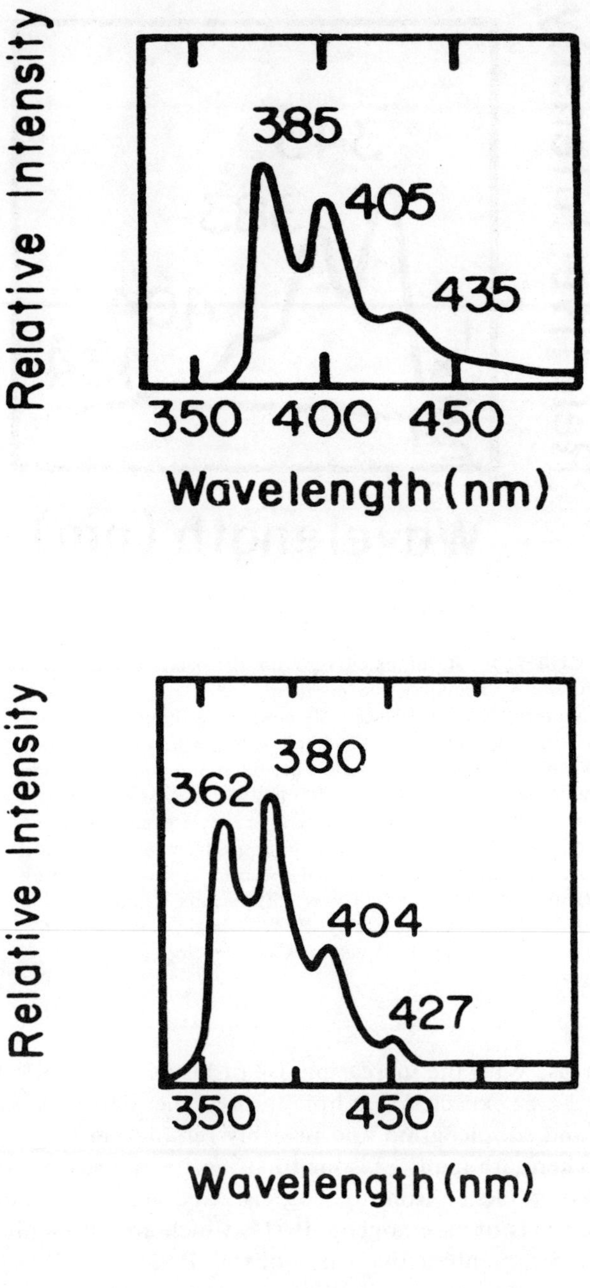

FIGURE 52. Fluorescence emission spectra of standard chrysene and benz(*a*)anthracene samples.
(a) Fluorescence emission spectrum of a standard sample of benz(*a*)anthracene (in methanol); λ_{exc} = 296 nm
(b) Fluorescence emission spectrum of a standard sample of chrysene (in methanol); λ_{exc} = 296 nm. (Reprinted with permission from Fox, M. A. and Staley, S. W., *Anal. Chem.*, 50, 967, 1978. Copyright by the American Chemical Society.)

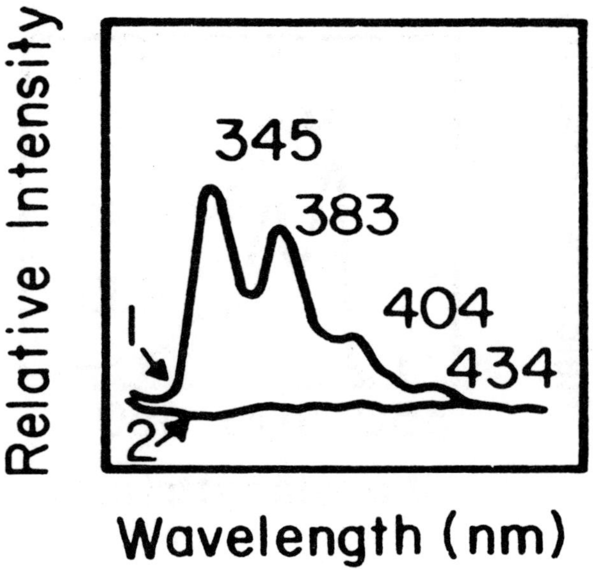

FIGURE 53. Resulting curves from the selective modulation procedure to analyze the mixture of chrysene and benz(*a*)-anthracene. Curve 1: Selectively modulated fluorescence emission spectrum of the same sample whose normal spectrum is shown in Figure 52; excitation modulation at the absorption maxima of chrysene. (w_{ex} = 2.0 mm, w_{em} = 0.5mm, 1.5nm/s, sens. = 3% full scale, J = 1 S). Curve 2: Selectively modulated (i.e., nulled) fluorescence emission spectrum of a standard chrysene sample (whose normal spectrum is shown in Figure 52) under conditions as in Curve 1. (Reprinted with permission from Fox, M. A. and Staley, S. W., *Anal. Chem.*, 50, 967, 1978. Copyright by the American Chemical Society.)

citation xenon sources. With the increasing use of lasers as excitation sources in fluorescence studies, it can be expected that hplc fluorescence detectors will use some form of laser excitation and subpicogram and possibly femtogram (10^{-15} g) levels of PAH will be measured. There are many reasons to strive for the realization of these lower detection limits. First, evidence from toxicologists and medical researchers is far from conclusive as to the levels of carcinogenic PAHs which are medically significant. Second, many of the more highly active carcinogenic PAH are of very high molecular weight, e.g., the dibenzanthracenes and dibenzopyrenes. The possible health hazards of even higher molecular weight PAH, i.e., MW > 300, are for the most part unknown and unreported in the literature. These PAH are only very slightly soluble in water and are for the most part undetected at present, even after preconcentration steps. If the concentrations of these PAH in water systems could be determined, more could be learned about the possible health effects of these compounds. Finally, most of the error and uncertainty present in the analyses of PAH in water arises from experimental problems in the extraction and preconcentration steps, and lower limits of detection would make it possible to forego this step. Since hplc reversed phase analyses are carried out with aqueous mobile phases, the water sample could be directly injected into the liquid chromatograph.

The major drawback of fluorescence detection is the inability of the method to provide qualitative as well as quantitative identification of peaks. Whereas mass spectrometry and infrared spectroscopy can provide qualitative information to allow identification of individual PAH, solution fluorescence spectra taken at room temperature are seldom sufficiently detailed to allow unambiguous identifications. Therefore, a standard sample of every component present is necessary in order to confirm the presence of a compound by fluorescence analysis. Matrix isolation techniques can be used to overcome this problem. Fluorescence spectra recorded at 77K with the PAH dispersed in an inert gas matrix are capable of providing fingerprint-like detail sufficient for identification. However, this technique is at present in the development stage and is not yet capable of on-line hplc monitoring. Development work is also in progress on hplc-ms interfaces[160] which should provide the same degree of information that is presently available from gc/ms systems. Before systems of this type are available to the analyst the direct introduction of an hplc eluent into a mass spectrometer must be resolved.

At present, sample cleanup and PAH fractionation are still essential prior to hplc analysis. Although model solutions can be readily handled, environmental samples are rarely simple. Since fluorescence detection requires the use of standards, the extent of interpretation of environmental samples is limited by the number of available standards. Coeluting peaks and interfering fluorescence spectra make one-step identifications by fluorescence techniques impossible.

To simplify the interpretation of the results of fluorescence analysis, many of the column and thin layer chromatographic techniques as well as the two stage hplc separation of Wise et al.[157] can be employed as a separation step. The degree of cleanup necessary obviously depends on the complexity of the sample. Fresh water lake samples would not be treated in the same way as an oil slick water sample.

Hplc has proven successful in resolving many isomeric PAH. Further investigations of C_{18} packings should proceed with a view to the development of a packing that will resolve a greater number of isomers. The continuing development of gradient elution techniques in conjunction with reversed phase hplc applications will enhance the use of hplc for PAH analyses.

V. GAS CHROMATOGRAPHY

A. Introduction

The required properties of a gc stationary phase for separation of PAH are selectivity for individual compounds and thermal stability over the range of temperatures necessary to elute all components. For packed column gas chromatography, no material has yet been developed that satisfies both these demands. Despite this drawback, packed column gc has been widely used over the past 20 years for analyses of PAH. Two major advances in the past several years have now led to a sharp decline in the use of packed column gc. First, the development of hplc stationary phases selective to PAH coupled with the increase in the availability of hplc gradient elution techniques has led to a shift away from packed column gc to the use of hplc. Second, a number of difficulties associated with the use of capillary columns in gc have been eliminated. With the increased efficiency of capillary columns, it is now possible to resolve complex mixtures of PAH with the same stationary phases that have been used in packed gc columns and at lower column temperatures. As an indication of this trend, PAH analyses by gc at Batelle's Columbus Laboratories are being done by capillary column gc. The present status and future outlook of capillary and packed column gc methods for analyses of PAH are reviewed.

Flame ionization detection is most widely used in PAH analysis. A number of specialized techniques such as gas phase fluorescence and photoionization are in developmental stages and matrix isolation fluorescence and Fourier Transform-IR methods may have future application. The combination of gas chromatography with mass spectrometry, gc/ms, is a powerful analytical method. The advantages of such a system especially when coupled with a computer data system make this detection system the first choice for PAH analysis.

B. Packed Column Separations

1. Introduction

When environmental or "real" samples are of interest, the laboratory equipped with capillary column gc or fully committed to its introduction should have limited need for packed columns. However, packed columns are being used successfully for analyses of PAH.

A number of stationary phases have been employed for PAH analyses; among these are SE-30, OV-1, OV-101, OV-7, and Dexsil® 300 and 400. All of these stationary phases have approximately the same selectivity for PAH. The Dexsil® phases show the best thermal stability on the basis of low column bleed at temperatures up to 320°C, therefore, packed column applications with non-Dexsil® stationary phases are not covered in this chapter.

A number of liquid crystal stationary phases exhibit remarkable selectivity towards individual PAH. However, their applicability is limited to small temperature regions between 200 and 300°C, and this severely limits the application of temperature programming. These stationary phases also exhibit high levels of column bleed and this results in columns with lifetimes of 1 month or less. Despite these serious disadvantages, these stationary phases can be useful for certain specific applications.

2. Practical Considerations

The selection of a carrier gas, flow rate, solid support, and temperature program are important parameters in packed column separations but are not the focus of current research. Helium and nitrogen are both equally suitable as carrier gases. For maximum efficiency and the prevention of column deterioration, provisions must be made to reduce the oxygen and water content of the carrier gases. Oxygen scrubbers and molecular sieve traps to remove moisture are commercially available.

The flow rate, u, of the carrier gas can have an effect on column efficiency and this is expressed in the van Deemter equation,[161] $H = A + B/U + Cu$ where A, B, and C are constants characteristic of the column, column solutes, stationary phase and mobile phase, and H is the height equivalent to a theoretical plate. Plots of H versus u should result in a curve with a minimum H value at the optimum flow rate.

The choice of a solid support is less important than that of a stationary phase but the support should be chemically inert toward all substances and should have a large surface area that results in uniform liquid phase coverage. The most commonly used support for Dexsil® phases is white diatomite derived from filter-aid and is commercially available as Chromosorb® W-HP manufactured by Johns-Manville.

Finally, one must incorporate an appropriate temperature program since PAH include compounds ranging from molecular weight, 128, naphthalene, to molecular weight 300, coronene. Retention times under isothermal conditions are dependent on the molecular weight and there is no single temperature at which an isothermal separation of all PAH is feasible. A temperature range from ca. 80 to 300°C is satisfactory for compounds from naphthalene to coronene. With temperature programming one obtains a chromatogram with narrow, evenly spaced peaks over the entire molecular

Table 33
CHARACTERISTICS OF DEXSIL STATIONARY PHASES

DEXSIL® 300 GC (carborane-methyl silicone) (50 to 400 + °C)	DEXSIL® 300, with proper conditioning and on a carefully prepared support (no trace base, or Fe), can be used to 400 + °C and is the most thermally stable stationary phase.
DEXSIL® 400 GC (carborane-methylphenyl silicone) (20-375°C)	DEXSIL® 400, is only slightly more polar than DEXSIL® 300 but is less viscous and more efficient.
DEXSIL® 410 GC (carborane-cyanoethyl methyl silicone) (20-360°C)	DEXSIL® 410 is the most polar of the carborane silicones.

Properties	DEXSIL® 300 GC	DEXSIL® 400 GC	DEXSIL® 410 GC
Solubility	Soluble in ether, dichloromethane, chloroform, aromatic solvents. Insoluble in water, alcohols.		
Physical appearance	Waxy, white solid melting point 30-38°C	Opaque, viscous fluid	Opaque, viscous fluid
Siloxane-carborane ratio	4 to 1	5 to 1	5 to 1
Stability	Stable to most chemicals except strong bases		
Average mol wt	16,000-20,000	12,000-16,000	9,000
Temperature minimum/maximum	50-450°C (500°C, limited time)	20-375°C	20-360°C

From technical staff, Gas-Thin Layer-Liquid Chromatography-Chemicals and Accessories, Catalog 18, 1976, Analabs, Inc., A Unit of Foxboro Analytical, North Haven, CT. With permission.

weight range and this is analogous to the result achieved by gradient elution with hplc. A temperature gradient of 4°C/min is commonly reported in the literature but little loss in performance results from a gradient of 6 or 8°C/min. An advantage of the latter gradients is a reduced analysis time.

3. Dexsil® 300 Separations

Dexsil® 300 as well as the similar stationary phases Dexsil® 400 and 410 is manufactured by Olin Corporation and is available through Analabs, Inc. The structure of the phases, shown in Figure 54, consists of metacarborane units (carborane icosahedral cage structures) connected by siloxane units.[162] The physical properties of the Dexsil® series are listed in Table 33. The most important feature is the high temperature stability which is unique among stationary phases used in gc determinations of PAH. As a consequence of this thermal stability the Dexsil® packed columns exhibit a very low bleed rate. After 72 hr of conditioning at 380°C, a mass spectrum of the background and column bleed indicated only a small number of fragments.[163]

Lao et al. have studied the PAH fraction from a wide variety of environmental samples such as airborne particulate matter, coal tar, and wood preservative sludge using packed column gc with Dexsil® 300 and 400.[164-166] With a packing of 6% Dexsil® 300 on 80/100 mesh Chromosorb® W-HP, 124 PAH present in air samples were identified with more than 100 compounds sufficiently resolved for quantitative measurement.[165] However, with this column isomeric PAH, e.g., B(a)P and B(e)P, anthracene and phenanthrene, and chrysene and triphenylene were not resolved. It was found, however, that 6% Dexsil® 400 packings provided a modest resolution of B(a)P from B(e)P and chrysene from benz(*a*)anthracene.[165] In further studies,[167] Lao's group

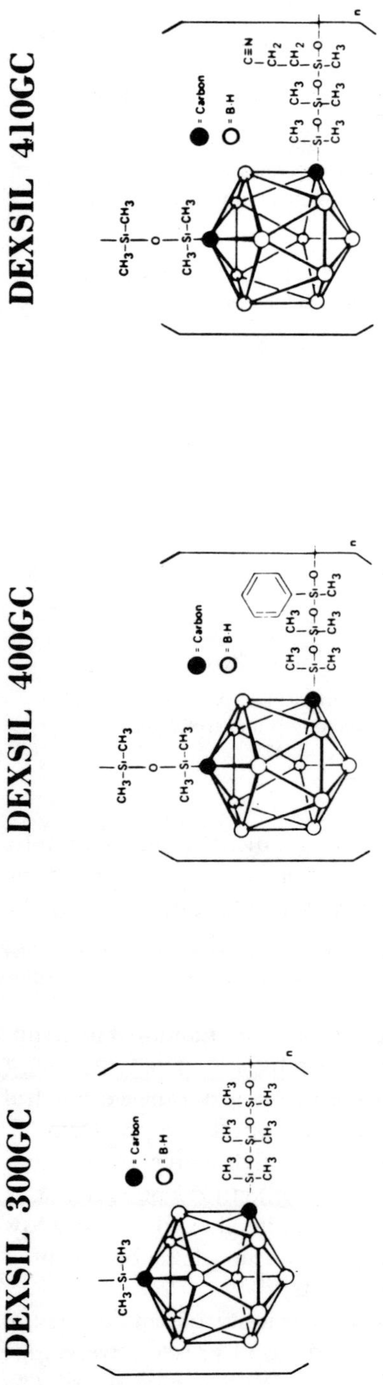

FIGURE 54. Structures of the Dexsil® series of gc stationary phases. (From technical staff, Gas-Thin Layer-Liquid Chromatography Chemicals and Accessories, Catalog 18, 1976, Analabs, Inc.; A unit of Foxboro Analytical, North Haven, Ct. With permission.)

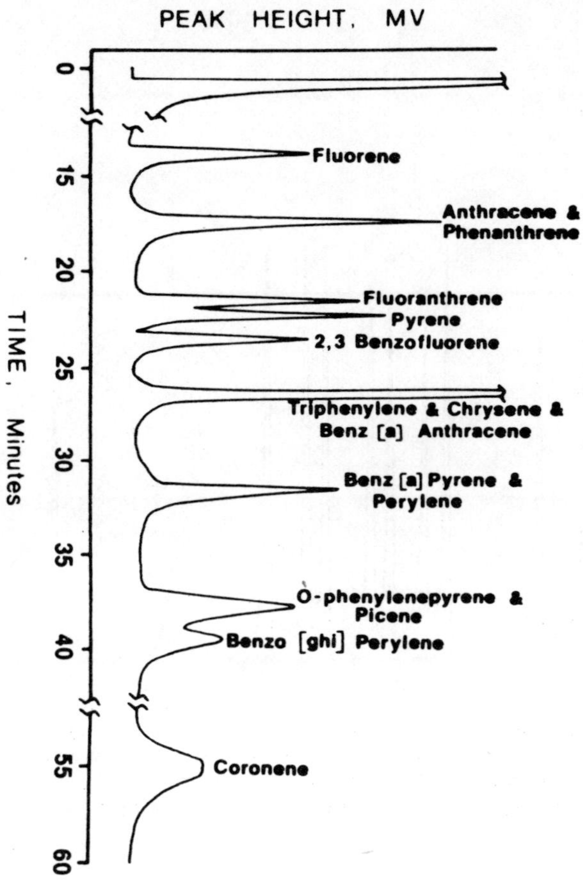

FIGURE 55. Chromatogram of 16 PAH mixture on Dexsil® 300. Column: 15 ft × 0.125 in. o.d. silanized glass. Packing: 3% (w/w) Dexsil® 300. Conditions: Oven 110 to 320°C, 4°C/min.; on-column injection; detector, 300°C; flow rate, 51 mℓ/min. Concentrations: 100 ppm. (Reprinted with permission from Strand, J. W. and Andren, A. W., *Anal. Chem.*, 50, 1508, 1978. Copyright by the American Chemical Society.)

studied various other environmental samples using 3% Dexsil® 300 packings and obtained similar results.

Strosher and Hodgson[168] determined PAH in waters and sediments from the Great Lakes after separation on a gc column packed with 3% Dexsil® 300 on 80/100 mesh Chromosorb® W/AW. The usual difficulties were encountered in the separation of isomeric PAH.

Strand and Andren[169] compared Dexsil® 300 to liquid crystal stationary phases for separations of PAH. Their separations on a 3% Dexsil® 300 column are shown in Figure 55. They point out that although the Dexsil® 300 packed column has the necessary thermal stability required to elute six and seven membered rings, it is incapable of resolving isomeric PAH.

In our studies of separations of PAH with a 3% Dexsil® 300 packed column, we have noted the excellent thermal stability necessary to elute higher molecular weight compounds such as coronene. A temperature gradient of 8°/min from 30 to 330°C was employed. The level of column bleed at 330°C was not serious. The complete resolution of an 18 component mixture of PAH is illustrated in Figure 56. The mixture

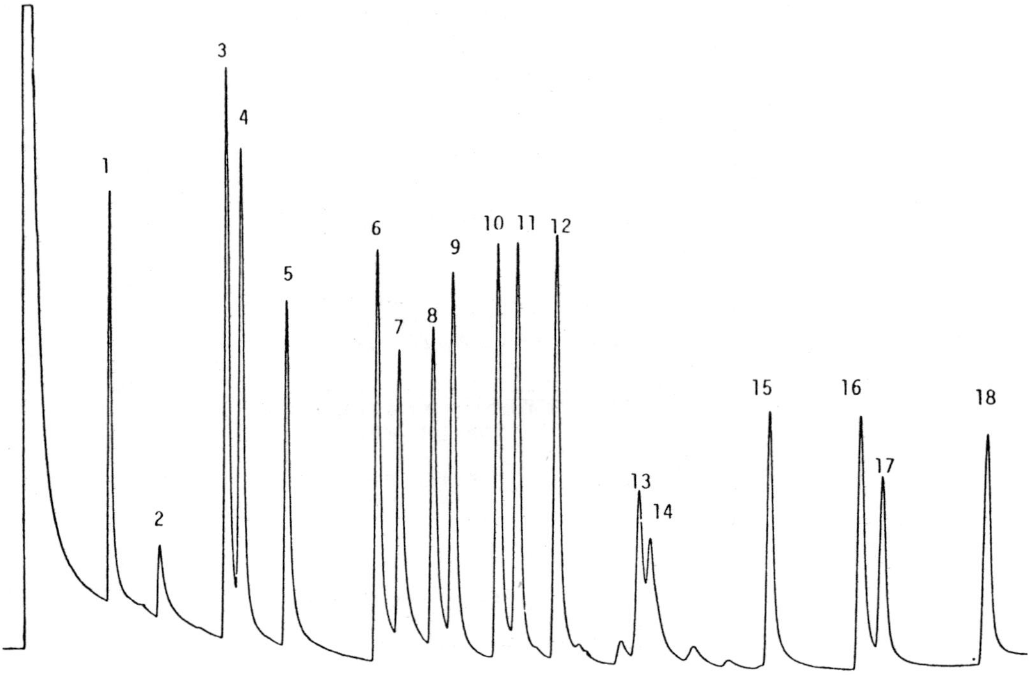

FIGURE 56. Gas chromatogram of a synthetic mixture of 18 PAH on a Dexsil® 300 packed column. Column: 6′ × 1′8″ o.d. stainless steel. Packing: 3% Dexsil® 300 gc on Chromosorb W-HP (100/120 mesh). Conditions: Initial oven temperature, 80°C; Program rate, 8°C/min; final temperature, 330°C, injector, 275° C; detector, 330°; flow rate, 35mℓ/min. He. Concentration ca. 100 ppm.

Legend

1.	Naphthalene	10.	Fluoranthrene
2.	2-Methylnaphthalene	11.	Pyrene
3.	Acenaphthylene	12.	Benzo (*b*) fluorene
4.	Acenaphthene	13.	Benz (*a*) anthracene
5.	Fluorene	14.	Naphthacene
6.	Phenanthrene	15.	Benzo (*a*) pyrene
7.	Carbazole	16.	Dibenz (*a, h*) anthracene
8.	2-Methylanthracene	17.	Benzo (*ghi*) perylene
9.	9-Methylanthracene	18.	Coronene

does not include all of the isomers at MW:228 and 252 which, if present, would coelute with B(a)A and B(a)P. However, a number of separations should be noted, e.g., acenaphthylene and acenaphthene, phenanthrene and carbazole, 2- and 9-methylanthracene, pyrene and fluoranthene, benz(*a*)anthracene and naphthacene, and dibenz(*a,h*)anthracene and benzo(*ghi*)perylene. In addition, PAH of molecular weight greater than 252, e.g., dibenz(*a,h*)anthracene and coronene are eluted as sharp peaks and coronene is eluted in less than 30 min. Retention times for the PAH studied are listed in Table 34. These values were reproducible to less than 1% within a single day and less than 2% over a period of months with the same column and operating parameters. From these results it can be seen that Dexsil® 300 is unsuitable as a gc column packing for the resolution of isomeric PAH. However, when the separation of the isomers at MW:228 and 252 is not important, Dexsil® 300 is a good packing since it has the necessary thermal stability to elute coronene in less than 30 min and will provide reproducible retention times for PAH ranging from naphthalene to coronene.

Table 34
RETENTION TIMES FOR 18 PAH STANDARD ON A 3%
DEXSIL® 300 GC COLUMN

Compound	Average retention time (sec)[a]	SD (sec)[b]
Naphthalene	151	3.7
2-Methylnaphthalene	236	4.1
Acenaphthylene	355	3.5
Acenaphthene	383	3.3
Fluorene	463	4.1
Phenanthrene	626	5.8
Carbazole	665	5.8
2-Methylanthracene	725	6.1
9-Methylanthracene	762	4.2
Fluoranthene	842	5.3
Pyrene	878	5.4
Benzo(*b*)fluorene	944	5.2
Benz(*a*)anthracene	1091	7.3
Naphthacene	1107	7.0
Benzo(*a*)pyrene	1324	6.4
Dibenz(*a,h*)anthracene	1480	5.8
Benzo(*ghi*)perylene	1516	6.0
Coronene	1700	7.8

Note: Chromatographic conditions identical to Figure 56.

[a] Average of five replicate runs.
[b] SD = standard deviation based on five runs.

4. Liquid Crystal Stationary Phases

The liquid phases developed by Janini et al.[170-173] at the National Cancer Institute in Maryland exhibit remarkable selectivity towards PAH including isomeric species. *N,N*′-bis(*p*-methoxybenzylidene)-α,α'-bi-*p*-toluidine, as well as the p-butoxy, p-hexyloxy, and p-phenyl analogs of the above have all been investigated as stationary phases for PAH separations. The structure of the methoxy liquid crystal is:

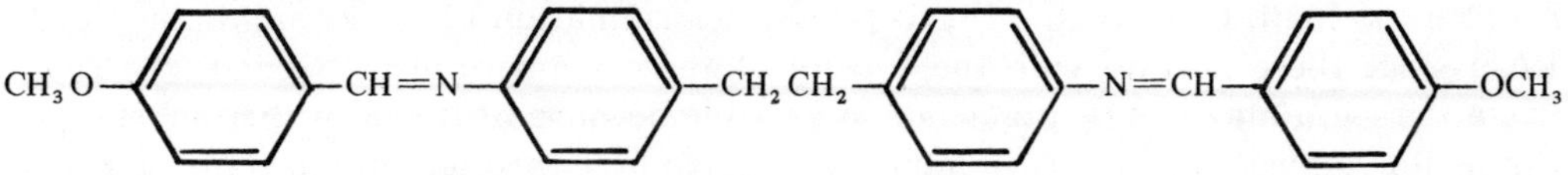

and the others are similar with the replacement of the methoxy groups by another group. The shorthand symbol, BMBT, is given to the methoxy liquid crystal with BBBT, BHxBT, and BPhBT used to denote the analogous versions.

Separation on the liquid crystal phase is based on one essential characteristic of the PAH. The various PAH, including the structural isomers, differ in their length to width ratio, or rod-like character. The more rod-like the PAH, the greater is the interaction between the PAH and the nematic solvent. As an example, anthracene is retained more strongly than phenanthrene thereby providing a separation of the two isomers.

These stationary phases have a number of practical limitations. First, separations occur only when the liquid crystal is in the nematic phase. In many cases, this phase exists in a narrow temperature region and this limits the effective application of temperature programming. In addition, the lower temperature limit of the nematic region

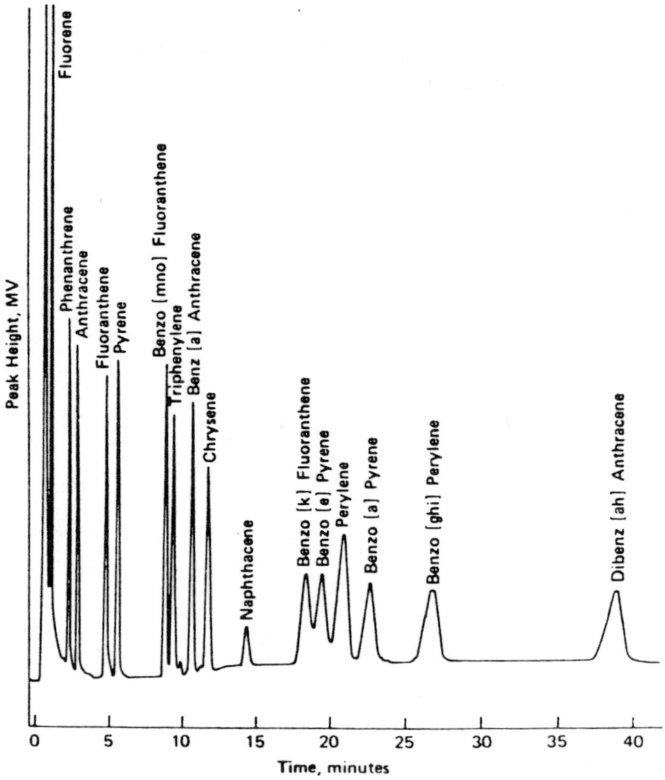

FIGURE 57. Chromatogram of a synthetic mixture of sixteen polycyclic aromatic hydrocarbons of wide molecular weight range. Column: 4 ft. × 0.125 in. o.d. stainless steel. Packing: 2.5% (w/w) BMBT. Conditions: Initial oven temperature, 185°C; initial hold, 2 min. Program rate, 4°C/min; final temperature, 265°C; injector, 265°C; detector, 265°C; flow rate, 40 mℓ/min. Concentration ranges from about 0.1 μg of phenanthrene to about 0.2 μg for benzo(*ghi*)perylene, except for naphthacene, which was less, but undetermined. (Reprinted with permission from Janini, G. M., Johnston, K., and Zielinski, W. L., Jr., *Anal. Chem.*, 47; 670, 1975. Copyright by the American Chemical Society.)

is often too high, (180 to 200°C), to permit determination of lower molecular weight PAH since these coelute with the solvent at these temperatures. Second, the entire nematic region may not be practically accessible because of the poor thermal stability of the liquid crystal phases. Column bleed is often high throughout the entire workable range and tends to become excessive at the higher temperatures. As a result of this poor thermal stability, retention times often change drastically over a period of weeks. Third, the efficiency of columns packed with liquid crystal phases is low, i.e., peak shapes are generally broad and this limits the total number of compounds which can be resolved. However, despite these serious disadvantages, the liquid crystal phases are the only packings truly able to differentiate between isomeric PAH at this time.

BMBT was the first liquid crystal to be studied by Janini et al.[170] The nematic region is from 181 to 320°C. However, after 72 hr of operation of the column at 300°C a loss of 10% of the stationary phase occurred; an upper temperature limit of 265° was considered necessary to reduce column bleed. The separation of a sixteen component PAH mixture on a 2.5% (w/w) BMBT-Chromosorb® W-HP column packing is illustrated in Figure 57. The separation of isomers is remarkable. The following isomeric PAH were baseline resolved: anthracene/phenanthrene, fluoranthene/pyrene, triphenylene/benz(*a*)anthracene/chrysene/naphthacene and benzo(*k*)fluoranthene/

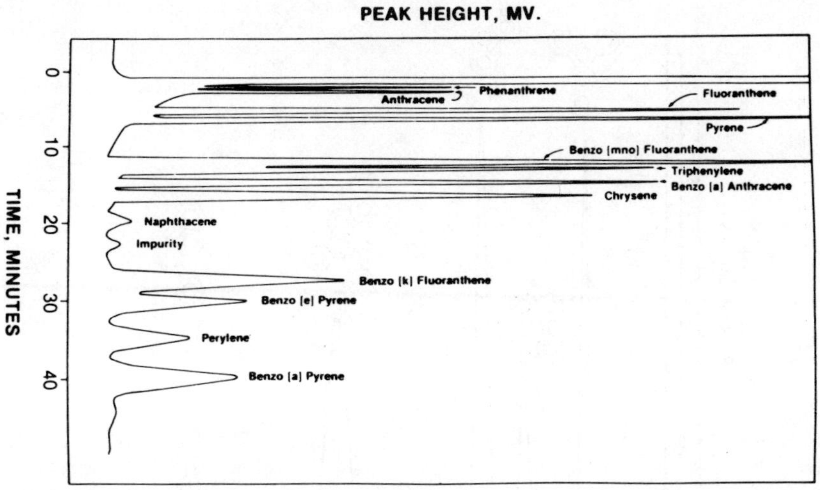

FIGURE 58. Chromatogram of a synthetic mixture of 13 polycyclic aromatic hydrocarbons of wide molecular weight range. Column: 6 ft. × 0.125 in. i.d. Supelco® glass. Packing: 2.5% (w/w) BBBT. Conditions: initial oven temperature, 210°C; initial hold, 3 min; program rate, 4°C/min; final temperature, 255°C; flow rate, 24 mℓ/min; electrometer setting, 16 × 10^{-11} A full scale. (Reprinted with permission from Janini, G. M., Muschik, G. M., and Zielinski, W. L., Jr., *Anal. Chem.*, 48, 809, 1976. Copyright by the American Chemical Society).

benzo(*a*)pyrene/benzo(*e*)pyrene/perylene. Since benzo(*k*)fluoranthene has the largest length to width ratio of all the benzofluoranthene isomers, others which may be present in environmental samples would be expected to elute earlier and not interfere with the benzopyrenes and perylene.

In an effort to increase column lifetimes by a reduction in column bleed, Janini et al. developed the p-butoxy analog, BBBT.[171] Whereas the BMBT column lost about 16% of its liquid phase content when operated at 260° for 170 continuous hours, the losses for BBBT column were sufficiently low that the HETP was reproducible to 3% as measured on the benz(*a*)anthracene peak. The nematic range of BBBT is 188 to 303°, but the effective upper temperature is 265°C. A typical separation of 13 PAH with BBBT is shown in Figure 58; these separations are as good as those achieved with BMBT shown in Figure 57.

For PAH with MW > 252. Janini et al.[172] prepared two new liquid crystals, BHxBT and BPhBT, the p-hexyloxy and p-phenyl analogs of BMBT. The theoretical plate count for chrysene was reproducible to within 5% after 100 hr of continuous operation at 275°C for the BPhBT and BHxBT phases. It was found that one-inch-long plugs of a preconditioned (5 to 6%) SE-30 packing at both ends of the column increased column lifetimes with no loss in performance. The separation of five 22-carbon PAH at 280°C under isothermal conditions are illustrated in Figure 59. The dibenzanthracene isomers were resolved from each other as well as from benzo(*ghi*)perylene and picene. Five 24-carbon PAH four dibenzopyrene isomers and coronene, were baseline resolved, (Figure 60), demonstrating that PAH of molecular weight over 300 can be separated on liquid crystal phases.

Strand and Andren[169] have also examined separations of PAH on BBBT and BPhBT stationary phases. Under operating conditions essentially identical to those used by Janini's group, the chromatograms of a 15 PAH mixture shown in Figures 61 and 62 were obtained. Separations were excellent but the efficiencies of the columns were poor compared to Janini's columns. It was noted that retention times increased by more than 10% when the BBBT column was operated continuously for 1 week at 190°C.

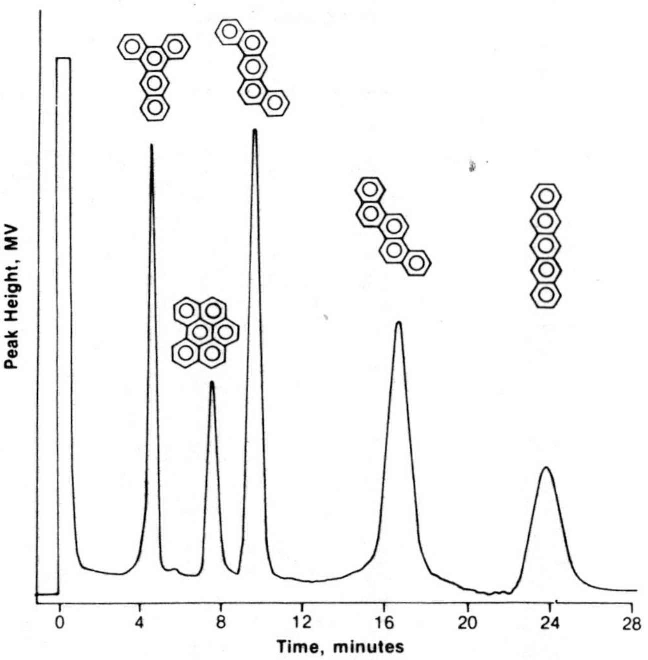

FIGURE 59. Chromatogram of five 22-carbon PAH on BPhBT at 270°C. Column: 2 ft, × 2 mm i.d. glass. Packing: 2.5% (w/w). Conditions: injector and detector, 280°C; flow rate, 30 mℓ/min. Compounds in order of elution are: 1,2,3,4-dibenzanthracene (dibenz(*a,c*)anthracene), 1,12-benzperylene (benzo(*ghi*)perylene), 1,2,5,6-dibenzanthracene (dibenz(*a,h*)anthracene), 1,2,7,8-dibenzphenanthrene (picene), and 2,3,6,7-dibenzanthracene (pentacene). (Reprinted with permission from Janini, G. M., Muschik, G. M., Schroer, J. A., and Zielinski, W. L., Jr., *Anal. Chem.*, 48, 1879, 1976. Copyright by the American Chemical Society.)

The BPhBT column maintained reproducible retention times for 1 month while in continuous operation at 270°C.

When hplc or capillary column gc methods are unavailable, packed column gc with liquid crystal stationary phases are a useful alternative. Unfortunately, no liquid crystal column is able to handle the entire molecular weight range of compounds present in an environmental sample. Therefore, two columns in series can be used, e.g., a Dexsil® 300 column followed by a BPhBT column. It is unfortunate that a class of stationary phases that is so highly selective to PAH should have serious limitations in terms of thermal stability, column efficiency, and wide molecular weight range applicability. Unless further research in this area overcomes these problems, it is highly unlikely that liquid crystal phases will be widely used.

C. Capillary Column Gas Chromatography

1. Introduction

The development of capillary column gc has had a significant impact on methods of analysis of environmental samples. Although introduced more than two decades ago, recent instrumental advances have made it possible to take greater advantage of this technique. Novotny[174] has reviewed the history, analytical methodology, and applications of the entire field of capillary gc. Grob and Grob[175] have thoroughly discussed the technical difficulties of introducing glass capillary gc into the laboratory.

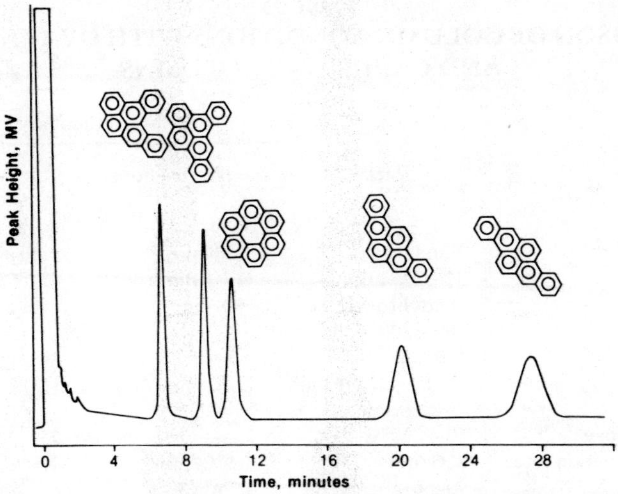

FIGURE 60. Chromatogram of five 24-carbon PAH on BPhBT at 290°C. Column: 2 ft × 2 mm i.d. glass. Packing: 2.5% (w/w) BPhBT. Conditions: injector and detector, 290°C; flow rate, 50 mℓ/min. Compounds in order of elution are: 4,5,6,7-dibenzpyrene (dibenzo(*def,p*)chrysene, 4,5,7,8-dibenzpyrene, coronene, 2,3,6,7-dibenzpyrene (benzo(*rst*-)pentaphene), and 1,2,6,7-dibenzpyrene. (Reprinted with permission from Janini, G. M., Muschik, G. M., Schroer, J. A., and Zielinski, W. L., Jr., *Anal. Chem.*, 48, 1879, 1976. Copyright by the American Chemical Society.)

The advantages and disadvantages of glass capillary gc and packed column gc as discussed by Grob and Grob[175] and Cram and Yang[176] are summarized here.

Glass capillary and packed column gc were compared by analyzing the same lake water extract by each method.[175] OV-1 was used as the stationary phase in both columns. Because Grob and Grob have been using glass capillary columns in their laboratory since the mid 1960s and are expert in their preparation, the capillary columns used in this comparative study were prepared in their own laboratory. To avoid bias, the packed columns were supplied by an expert not associated with their laboratory. The analytical conditions for the packed and capillary columns are listed in Table 35 and the chromatograms obtained from the two columns are shown in Figure 63.

The capillary column was 12 times longer than the packed column, and contained less than 2% of the stationary phase of the packed column. The separation efficiency was higher for the capillary; 490 peaks were observed as compared to 118 for the packed columns. It should be noted that it required many trials to achieve this degree of separation with the packed column whereas the capillary column separation was "routine".[175]

The advantages of glass capillary gc have been listed by Cram and Yang.[176] The glass capillary columns perform more reproducibly than packed columns and baseline stability and resolution are excellent. The peaks obtained with capillary columns are narrower, with essentially triangular symmetry which facilitates quantitation. Since the stationary phase is not coated on a support, the permeability of the capillary column is very high so that longer columns can be used. This added length increases the efficiency of the capillary column and even with this added length the time of analysis is reduced. Because of the increased efficiency, the choice of stationary phase is in general not as critical as for packed columns, but in spite of this, several isomeric PAH have yet to be resolved.

Table 35
COMPARISON OF COLUMN CONDITIONS WITH OV-1 IN PACKED AND CAPILLARY COLUMNS

		Capillary	
	Packed	Narrow bore	Wide bore
Length	3 m	35 m	60 m
Inner diameter	2 mm	0.28 mm	0.60 mm
Coating	3% OV-1 on Gaschrom Q®, 80—100 mesh	OV-1 film 1.0 × 10^{-5} cm	OV-1 film 1.0 × 10^{-5} cm
Liquid-phase-load	120 mg	1.5 mg	14 mg
Carrier gas flow	He 10—22 mℓ/min (programmed)	H_2 2.2 mℓ/min	H_2 14 mℓ/min
Temperature range	50—200°C	25—170°C	25—200°C
Temperature program rate	2°C/min	3.5°C/min	2.5°C/min
Analysis time	90 min	45 min	65 min
Sample size	0.15 μℓ direct	0.6 μℓ splitting 1:25	0.6 μℓ splitting 1:8
Attenuation	× 32	× 4	× 16
Number of peaks	118	490	320

From Grob, K. and Grob, G., *Identification and Analysis of Organic Pollutants in Water,* Keith, L. H., Ed., Ann Arbor Science Publishers, Ann Arbor, MI, 1975. With permission.

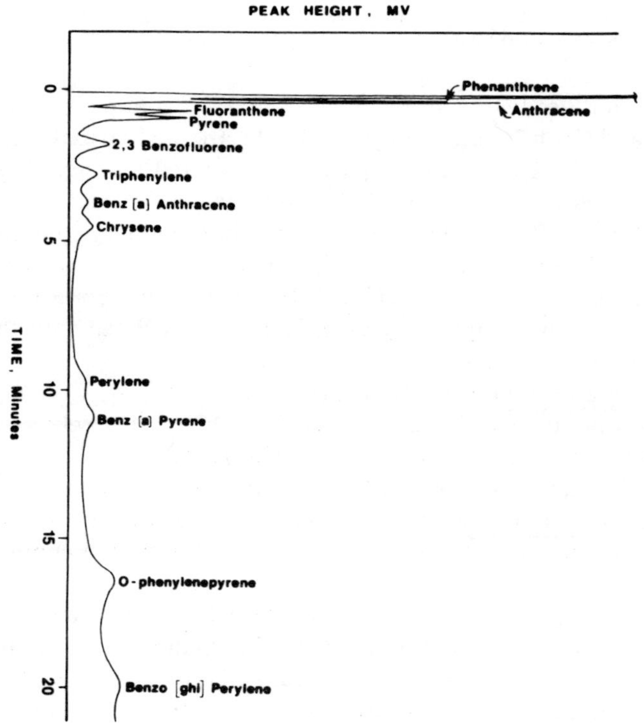

FIGURE 61. Chromatogram of 15 PAH mixture on BBBT. Column: 4 ft. × 0.125 in. o.d. silanized glass. Packing: 2.5% (w/w) BBBT. Conditions: Oven 190-270°C, 2-min hold, 4°C/min; on-column injection; detector, 250°C; flow rate, 40 mℓ/min. Concentrations: 100 ppm. (Reprinted with permission from Strand, J. W. and Andren, A. W., *Anal. Chem.,* 50, 1508, 1978. Copyright by the American Chemical Society.)

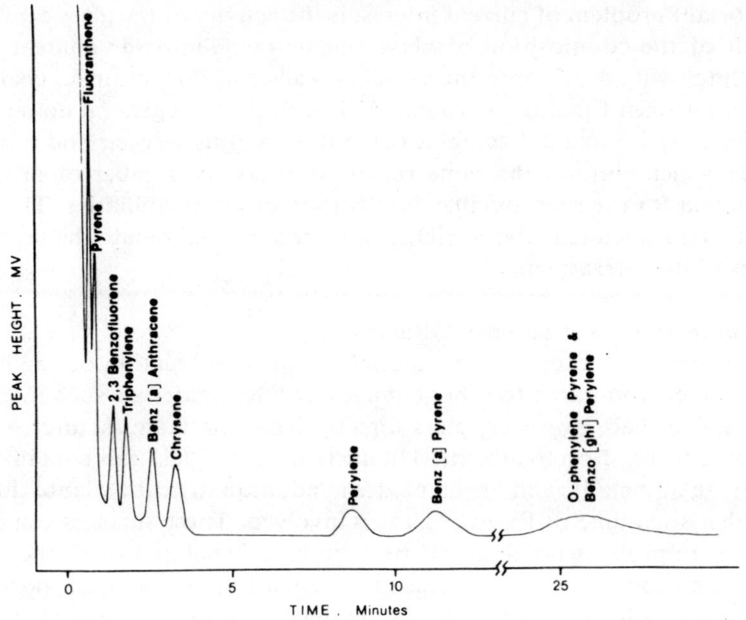

FIGURE 62. Chromatogram of 15 PAH mixture on BPhBT. Column: 6 ft. × 0.125 in. o.d. silanized glass. Packing: 2.5% (w/w) BPhBT. Conditions: Oven 270°C; on-column injection; detector, 250°C; flow rate, 20 mℓ/min. Concentrations: 100 ppm. (Reproduced with permission from Strand, J. W. and Andren, A. W., *Anal. Chem.*, 50, 1508, 1978. Copyright by the American Chemical Society.)

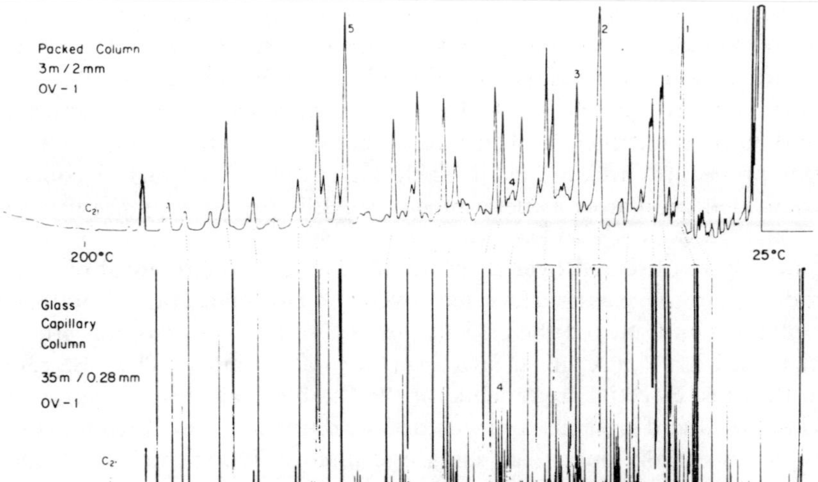

FIGURE 63. Comparison of typical separation from packed and capillary columns. Sample (extract from river water) and liquid phase identical for both columns. For column characteristics and operating chromatographic parameters, see Table 35. Dotted lines indicate corresponding sample components. (From Grob, K. and Grob, G., *Identification and Analysis of Organic Pollutants in Water*, Keith, L. H., Ed., Ann Arbor Science Publishers, Ann Arbor, MI, 1975, 75. With permission.)

An important problem of current interest is the activity of the glass capillary surface. As a result of the composition of glass and its metallic oxide content, many polar organic solutes will adsorb onto the capillary walls and this results in peak tailing with oxygen and nitrogen functional groups. Although PAH in general do not suffer from this problem, environmental samples contain numerous oxygen and nitrogen substituted PAH which elute in the same region as PAH. A number of procedures have been developed for the reproducible deactivation of glass capillaries. The recent introduction of fused silica (ca. 100% SiO_2) capillaries may eliminate the need for exhaustive deactivation pretreatments.

2. Deactivation of Glass Capillary Columns

Glass is generally chosen as the material for the construction of capillary columns because it is inert compared to other common column materials such as copper, stainless steel, and nickel. However, glass directly from the manufacturer is not suitable without some form of pretreatment. The surface of the glass can contain an appreciable percentage of metallic and boron oxides in addition to acidic silanol sites depending upon whether soda-lime or Pyrex® glass is involved. These surfaces can be quite "active" leading to problems such as: (1) peak tailing, (2) solute losses, and (3) stationary phase decomposition. These problems are particularly severe when polar solutes are analyzed on nonpolar stationary phases and when higher column temperatures necessary for high molecular weight compounds are used.

Some of the methods which have been developed to deactivate glass surfaces include silanization, pretreatment with thin-layers of nonextractable polymers, and acid leaching. The degree of column deactivation is usually judged by the chromatographic performance on test mixtures of solutes of widely ranging polarity. Recently,[177] Auger Electron Spectroscopy, AES, was employed in an effort to relate elemental surface concentrations on the glass capillary to the degree of the column activity as judged by chromatographic performance.

Lee et al.[177] have studied the effects of various acid leaching procedures for deactivating the surfaces of glass capillaries. Two basic types of glass were examined, a Kimble® R6 Flint glass (soda lime), and a Corning Pyrex® (7740) borosilicate glass. The bulk compositions of the glasses are listed in Table 36. The capillary columns were drawn by a Shimadzu® Model GDM-1 glass drawing machine to an internal diameter of 0.28 to 0.30 mm and then broken into 4″ lengths prior to treatment. The various deactivation procedures are outlined in Table 37. The elemental surface concentrations of the deactivated capillaries as determined by AES spectroscopy are listed in Table 38.

Lee et al.[177] have observed over a period of years that the removal of metallic and boron oxides from the glass surface improved the chromatographic performance of capillary columns and gauged the effectiveness of various deactivation procedures by the degree to which a "pure silica, SiO_2" surface was obtained. The AES results indicated that the most effective treatments for the Kimble® glass were hydrochloric acid leaching at 100°C and gaseous hydrochloric acid/formic acid leaching. In the first procedure, an essentially pure silica surface resulted, while in the second, the sodium and calcium oxide contents were reduced to near zero. The boron content of Corning Pyrex® was also reduced to "zero" by leaching with hydrochloric acid at 100°C. Chromatograms of C_5 through C_{12} n-alcohols obtained on a nonpolar OV-101 Pyrex® capillary column deactivated in this manner displayed a low degree of peak tailing.[177] Samples containing ketones, amines, and free sterols were also successfully analyzed. In addition, this column exhibited excellent thermal stability, i.e., low bleed up to a temperature of 300°C.

Table 36
BULK GLASS COMPOSITIONS

	% Composition as oxide	
Elements	Kimble® R6 Flint	Corning Pyrex® 7740
Si	67.7	81
Na	15.6	4
Ca	5.7	0.5
Mg	3.9	—
Al	2.8	2
Ba	2.0	—
B	1.5	13
K	0.6	—

From Lee, M. L., Vassilaros, D. L., Phillips, L. V., Hercules, D. M., Asumaya, H., Jorgenson, J. W., Maskarinec, M. P., and Novotny, M., *Anal. Lett.*, 12(A2), 191, 1979. With permission.

Lee et al.[177] noted that difficulties exist in the deactivation process. First, the reproducibility of the various procedures were found to vary in an unpredictable manner. They suggest that investigation of the deactivation of surface silanol groups is necessary and that surface coating procedures might be useful in this respect. Second, it was found that a high percentage (ca. 20 to 30%) of the elemental surface content was adsorbed carbon which interferes with the deactivation process. They pointed out the need to prevent this carbon contamination but noted the difficulty of accomplishing this in a laboratory environment. They did not discuss the important question of stationary phase stability before and after deactivation.

Meuser et al.[178] at Battelle's Columbus Laboratories have also developed an acid leaching deactivation procedure for the preparation of thermally stable high temperature glass capillaries. The leaching was accomplished with 20% hydrochloric acid at 180°C for 16 hr under reduced pressure. Following this step, the acid was washed out and the column heated to 280°C at reduced pressure for 24 hr. Silylation with a 20% solution of diphenyltetramethylsilazone at 400° for 8 hr completed the treatment. Glass capillary columns treated in this manner and coated with SE-52 were used at temperatures up to 320°C for PAH separations. No results on column bleed levels were given.

Active glass capillary surfaces can catalyze the decomposition of the stationary phase at the high column temperatures necessary to elute the high molecular weight PAH. Schomburg et al.[179-181] have investigated this problem by focusing on three parameters which affect the level of column bleed; (1) type of glass (alkali or borosilicate), (2) surface properties after deactivation and roughening treatments, and (3) type and origin of the methylsilicone stationary phases.

The thermal stability of the capillary columns over extended time periods at high temperatures must be assessed on the basis of several factors: (1) capacity ratios of selected standard compounds, (2) separation efficiency in terms of theoretical plates per meter, (3) extent of surface deactivation as gauged by tailing measurements, and (4) polarity in terms of retention indices of standard compounds.[179] For their studies, borosilicate and alkali capillary columns of similar length and internal diameter coated with OV-101 and SE-30 were used. These columns were preconditioned by operating at 280° for about 12 hr at carrier gas flow rates of 1 mℓ/min. The bleed levels were measured as a function of temperature from 200 to 280°C. Details of the determination

Table 37
EXPERIMENTAL TREATMENT CONDITIONS

Column no.	Glass type	Pretreatment	Leach solution[a]
1	Kimble® R6	—	88% Formic acid
2	Kimble® R6	—	19% Hydrochloric acid
3	Kimble® R6	—	Distilled water
4	Kimble® R6	—	19% Hydrochloric acid (100°C)
5	Kimble® R6	1. Helium purge at 300°C for 24 hr 2. Gaseous hydrochloric acid for 1 hr at 300°C 3. Oven heat at 350°C for 2 hr.	88% Formic acid
6	Kimble® R6	1. Identical to column 5	19% Hydrochloric acid
7	Kimble® R6	1. Identical to column 5	Distilled water
8	Corning Pyrex®	—	19% Hydrochloric acid
9	Corning Pyrex®	—	19% Hydrochloric acid (100°C)

Note: Unless indicated in parenthesis, the wash was done at room temperature (25°C).

[a] 8 mℓ of leach solution were used under nitrogen pressure, for a period of 27 hr. This was followed by a 2 mℓ distilled water rinse and subsequent drying with helium.

From Lee, M. L., Vassilaros, D. L., Phillips, L. V., Hercules, D. M., Asumaya, H., Jorgenson, J. W., Maskarinec, M. P., and Noyotny, M., *Anal. Lett.*, 12(A2), 191, 1979. With permission.

of the mass of bleed components are found in Reference 179. The results of the bleed tests indicated that SE-30 was more stable towards catalytic decomposition than OV-101. The most important conclusion is that glass capillary columns made from alkali glass exhibit much higher levels of bleed than borosilicate glasses, regardless of pretreatment and deactivation procedures.

Treatment of the alkali glass columns with gaseous hydrogen chloride and hydrogen fluoride converts the alkali and alkaline earth metal oxides (K_2O, Na_2O, BaO, CaO) into the corresponding halides which remain on the glass, whereas on pretreatment with hydrochloric acid the oxides are also converted to the halides but are removed from the columns during the washing step. It was noted[179] that a decrease in the sodium oxide level decreased the catalytic decomposition of methyl silicones. Methyl silicones coated onto the unpretreated borosilicate columns showed very low levels of bleed, although the borosilicate glass contains a small percentage, ca. 4%, of alkali metal oxides. However, alkali and barium carbonate, coated borosilicate columns exhibited strong bleed and this suggests that the influence of the alkali metal oxides depends on the manner in which the oxides are chemically incorporated into the glass structure. One difficulty with the borosilicate capillary columns is the adhesion of the stationary phase film to the glass. The deposition of a layer of barium carbonate, silica, or sodium chloride greatly enhances the ability of the stationary phase to cover the glass capillary surface completely thus minimizing the effect of the active glass surface on column bleed. In terms of thermal stability of the stationary phase, the borosilicate columns with or without the deposition of a prelayer, were considered a better choice than soda-lime columns.

In the following article, Schomburg et al.[181] extended their studies to consider the effect of various deactivation procedures on the extent of tailing. The polarity mixture which was studied included C_{10}, C_{11}, and C_{12} paraffins, C_{10}, C_{11}, and C_{12} esters, n-octanol, 2,6-dimethylphenol, 2,6-dimethylaniline, and dicyclohexylamine. Soda-lime glass capillary columns were used for the experiments. Simple baking of the methylpolysiloxane stationary phase OV-101 at 350°C for 5 hr on an untreated soda-lime glass surface resulted in fairly good deactivation, i.e., minimal peak tailing up to ca. 220°C, however the thermal stability of the stationary phase was poor beyond this point. When the baking treatment was preceded by a dealkalinization step with hydrogen chloride gas, the thermal stability of the OV-101 was improved, but tailing of the various solutes became severe. The optimum deactivation procedure involved dealkalinization followed by passage of a dilute solution of hydrogen fluoride gas in nitrogen (0.1 vol. %) followed by a 30 min baking at 450°C. This deactivation procedure was also found to work well for an alkylpolysiloxane stationary phase containing carborane groups, e.g., Dexsil® 400. Excellent peak symmetry was achieved for solutes containing NH-groups or more than one heteroatom.

The effect on surface activity resulting from the deposition of Carbowax® 20M in glass capillary columns was investigated by Blomberg and Wannman.[182] For this series of tests Pyrex® and AR-glass® (a soda-lime glass) tubing was cleaned in chromic acid-sulfuric acid, rinsed with water and acetone, dried, and drawn into 20 m × 0.6 mm o.d. capillaries. Some of the capillaries were etched with gaseous hydrogen fluoride and others with hydrochloric acid. After pretreatment, the capillaries were coated dynamically (at 20 mm/sec.) with a solution of Carbowax® 20M (5% w/v) in dry dichloromethane. After coating, the solvent was evaporated. The ends of the capillary were sealed; some of the capillaries were baked for 24 hr at 280°C and then extracted with 16 mℓ of dichloromethane. A final coating of either Carbowax® 20M or a methylsilicone stationary phase, SP-2100, was placed on the capillary.

All the capillaries showed some degree of deactivation as a result of simple rinsing with dilute Carbowax® 20M solution. The hydrogen fluoride-etched Pyrex® and hydrochloric acid-etched AR-glass® capillaries showed some residual activity towards alcohols, e.g., 1-octanol was not eluted from either column. However, AR-glass® etched with hydrogen fluoride exhibited a lesser degree of activity in that the eluted 1-octanol exhibited severe tailing.

Heat treatment significantly improved the degree of deactivation of all the Carbowax® 20M treated capillaries; 1-octanol, dodecane, and naphthalene were eluted and gave symmetrical nontailing peaks.

Blomberg and Wannman[182] showed that the etching treatments were necessary to improve the wettability of the glass surface with Carbowax®. Coating of nonetched glass capillaries with Carbowax® resulted in the formation of lenses of thicker film on the capillary. However, after baking and extraction of the unetched capillaries, the degree of deactivation was acceptable to the authors.[182] With Pyrex® capillaries, etching increased the degree of deactivation whereas with AR-glass® the deactivation was slightly impaired.

For capillaries preheated with a thin nonextractable Carbowax® layer, thermal stabilities were similar for the differently pretreated columns. When SP-2100 was coated onto these capillaries, the AR-glass® columns showed a drastically increased rate of bleed, while the analogous Pyrex®-glass columns exhibited a bleed rate comparable to capillaries coated with only Carbowax®. The authors[182] suggested that the active AR-glass® surface combines with a large fraction of the Carbowax® functional groups, and this yields an inactive Carbowax® surface to which the SP-2100 adheres poorly.

Dandeneau et al.[183] have recently reported on the use of fused silica capillary columns. The columns are fabricated from amorphous silica (SiO_2) of greater than 99.9% purity. "Fused silica" is made by fusing synthetic silica produced by the combustion of silane or silicon tetrachloride in oxygen. The material prepared in this manner is essentially pure silicon dioxide that contains metal oxide impurities in the ppm range. This synthetic material contains fewer metal oxide impurities than natural quartz.

Preliminary studies by Dandeneau's group indicated that the fused silica surface was well suited to derivatization with the whole array of silanizing agents. Surface modification via specific coupling agents or silylation to increase the operating of the columns is thus possible.

A "polarity mixture" consisting of straight chain alcohols, cyclic and aromatic amines, alkanes, and phenols was used to test column performance. An unheated 22 m × 0.20 mm fused silica column was deactivated with Carbowax® 20M and coated with SP-2100. For all the compounds present in 1 to 2 ng amounts, peak tailing was minimal with this column. Test mixtures containing primary amines, complex alcohols and PAH also showed insignificant tailing. A mixture of underivatized volatile fatty acids was separated on a fused silica column deactivated with Carbowax® 20M and further coated with Carbowax® 20M and showed no tailing. Since Dandaneau et al.[183] have shown that fused silica columns are not completely inactive, some deactivation is usually necessary to eliminate residual activity, mainly of a Lewis acid nature. Preliminary studies by the authors[183] indicate that the temperature stability of these columns is good and low levels of column bleed are observed for an SP-2100 column at 280°C.

It is difficult to fabricate completely inactive capillary columns which also provide surfaces that are suitable for thorough stationary phase coating. A concentrated effort must be made to perfect deactivation procedures which will result in excellent stationary phase stability at temperatures of 300°C or above. Minimal solute peak tailing must also be achieved. The procedures should be equally applicable for borosilicate or alkali glasses. At the present time fused silica capillary columns provide the most inactive surface available.

3. Separations of PAH

We have already emphasized the importance of resolving the numerous structural isomers of the PAH fraction. State of the art capillary column gc is the best approach at this time, but not perfect. Research is being conducted in order to develop capillary columns capable of complete resolution of all PAH isomers. The resolution of PAH can be improved by development of either: (1) more highly efficient columns, or (2) more highly selective stationary phases. Most research effort falls in the first category although development of PAH selective liquid crystal stationary phases for capillaries is also in progress. High separation efficiency, column deactivation, and thermal stability of the stationary phase are all factors which must be considered.

A system for identifying the hundreds of peaks which may be found in a capillary chromatogram is necessary. A well-defined procedure for determining retention indices for PAH can aid in identification of compounds in unknown mixtures and can be useful in interlaboratory data comparisons. The need for such comparisons is even more critical when a gc/ms/data system is unavailable for positive peak identifications or standard samples of individual PAH cannot be obtained.

Lee et al.[184] have addressed the problem of establishing a set of retention indices for PAH for temperature-programmed capillary column gc. The Kovats index system which determines retention indices relative to a series of homologous n-hydrocarbons has shortcomings when applied to PAH. Retention data are reproducible only with identical chromatographic columns, isothermal conditions, and operating parameters. Lee et al. pointed out additional factors which affect the reproducibility of retention data. These include: (1) batch to batch differences in the chemical composition of the stationary phase, (2) reactions and decomposition of the stationary phase, (3) the de-

Table 38
CAPILLARY ELEMENTAL SURFACE CONCENTRATION AFTER TREATMENT

	% Surface concentration						
Column no.	Si	C	O	Na	Ca	Cl	B
1	7	27	52	11	3	—	—
2	9	23	54	11	2	0.5	—
3	8	26	51	11	3	—	—
4	14	29	58	—	—	—	—
4[a]	18	18	64	—	—	—	—
5	12	27	55	4	1	1	—
6	9	35	44	9	1	2	—
7	9	33	45	10	1	2	—
8	11	26	62	—	—	1	2
9	15	23	63	—	—	—	—
9[a]	14	28	58	—	—	—	—

[a] Indicated columns were heat treated at 250°C for 12 hr under helium carrier gas flow and analyzed again by AES.

From Lee, M. L., Vassilaros, D. L., Phillips, L. V., Hercules, D. M., Asumaya, H., Jorgenson, J. W., Maskarinec, M. P., and Novotny, M., *Anal. Lett.*, 12(A2), 191, 1979. With permission.

gree of deactivation of the column material, (4) the type of column material, e.g., alkali, borosilicate glass, or fused silica, and (5) the type of temperature program employed. More reproducible retention data is obtained when the sample is chemically similar to the series of standard reference compounds. The need for a chemically homologous series to serve as reference compounds becomes even more critical when temperature programming is used since the retention time depends on temperature. In temperature-programmed gc, the approximate expression relating the retention index to retention time is

$$I = 100 \frac{T_{R\ (substance)} - T_{R(C_Z)}}{T_{R\ (C_{Z+1})} - T_{R(C_Z)}} + 100Z \qquad (10)$$

where $T_{R\ (substance)}$ is the retention time of the compound for which the retention index is to be determined, $T_{R(C_Z)}$ and $T_{R(C_{z+1})}$ are the retention times for the reference standards, usually n-alkanes, which bracket the compound of interest and Z is the number of carbon atoms in the n-alkane standard that elutes just prior to the compound of interest.

For PAH analysis, a new set of standards (naphthalene, phenanthrene, chrysene, and picene) was chosen instead of the n-alkanes. Retention indices for over 200 compounds were measured. The actual indices were determined by analyzing the PAH with the four standard reference compounds on SE-52 coated glass capillary columns. SE-52 was selected because capillary columns coated with this stationary phase are highly efficient.[184] The columns were drawn from Kimble® R6 soda lime glass capillary tubing and were treated according to the following procedure: hydrochloric acid gas phase etch, formic acid wash, HMDS/TMCS gas phase silylation, and static coating. A 12 m × 0.30 mm i.d. capillary column with a film thickness of 0.34 μm and a 12 m × 0.28 mm i.d. capillary with a 0.17 μm film thickness were found to be equally efficient in terms of separation efficiency, thermal stability, and inertness. The oven temperature was programmed from 50 to 250°C at a rate of 2°C/minute for each run.

The indices were calculated from Equation 10 in the same manner as when n-alkane standards are employed, however, in this case the Z stands for the number of rings in the PAH standard that elutes prior to the solute of interest. A necessary requirement is that all four standards be eluted during the temperature programming period.

The film thickness does not affect the value of the retention indices. Excellent reproducibility of the values of the PAH retention indices was obtained over a period of several months of repeated column use. The retention indices for over 200 compounds are listed in Table 39. To test the reliability of the retention indices as a function of different temperature programming rates, 10 chromatographic runs were made. Three runs were at 4°C/min and seven runs were at 2°C/min. The average 95% confidence limits for the series of PAH retention indices calculated from these runs were ± 0.16 index units.

The development of this retention index system is relatively recent and has not yet been tested by other laboratories. The importance of standardization of retention data for complex mixtures of PAH is obvious. We think that present research in the field of PAH separations with capillary-column gc will eventually lead to the universal adoption of such a system. However, at present, various researchers have individual preferences for different stationary phases, stationary phase concentrations, column materials, column lengths, deactivation procedures, etc. Until the optimum parameters have been established little progress in the development of a standardized retention index system can be made.

The actual separations of PAH and chromatographic results obtained by Lee et al.[184] can be assessed by inspection of Figure 64 and Table 39. The efficiency of the capillary columns was approximately 3000 plates per meter, and the peak width at half height averaged over all of the PAH was 0.80 index units. Two PAH differing by one retention index unit were stated to have better than 50% area resolution. The separation of the majority of PAH shown in Figure 64 was good, although certain isomers remained unresolved, e.g., chrysene/triphenylene, the dibenzanthracenes, and also many of the alkyl isomers listed in Table 39. It can be seen (Figure 64), that some of the early eluting solutes exhibit peak tailing.

Other experimental results obtained by capillary column analyses of PAH should be examined relative to the work of Lee et al.[184] in an effort to find improvements in: (1) column efficiency, (2) stationary phase selectivity towards PAH, (3) stationary phase thermal stability, and (4) reproducibility of retention data. In addition, parameters which improve peak symmetry and decrease tailing must be determined and optimized.

Bjørseth et al.[185-188] have analyzed numerous environmental samples for PAHs using capillary column gc. Bjørseth considered SE-54 to be a better stationary phase than OV-1, OV-101, and OV-61 on the basis of separation efficiency, column bleeding, and long-term stability. The chromatographic conditions for analysis of PAH in various environmental extracts were as follows: glass capillary column, 50 m × 0.34 mm i.d. coated with SE-54, hydrogen carrier gas at ca. 3 mℓ/min at ambient temperature, injector, and detector temperatures, 275°C, initial column temperature, 115°C, programmed temperature rate, 3°C/min, final temperature, 260°C. Since these results were reported in the literature, Bjørseth has moved to Battelle's Columbus Laboratories and the deactivation procedure reported by Meuser et al.,[178] is most likely the preferred procedure. The separations which were obtained on an extract from marine sediments containing PAH are illustrated in Figure 65. The isomers: anthracene/phenanthrene, chrysene/benz(*a*)anthracene, and benzo(*e*)pyrene/benzo(*a*)pyrene were resolved. However, the same difficulties that were encountered by Lee et al.[184] were

Table 39
PAH RETENTION INDICES

Compound no.	Compound name	Average index	Standard deviation
1	1,2-Dihydronaphthalene	197.01	0.07[c]
2	1,4-Dihydronaphthalene	197.01	0.07[c]
3	Tetralin	197.04	0.05[d]
4	Naphthalene	200.00	
5	Benzo(*b*)thiophene	201.47	0.05[f]
6	Indoline	204.74	0.04[c]
7	Indole	205.26	0.08[d]
8	Quinoline	209.70	—[b]
9	Isoquinoline	215.61	1.01[d]
10	2-Methylnaphthalene	218.14	0.28[h]
11	2-Methylbenzo(*b*)-thiophene	218.74	0.04[c]
12	Azulene	219.95	0.24[c]
13	Quinoxaline	220.37	0.05[d]
14	3-Methylbenzo(*b*)-thiophene	221.02	0.03[c]
15	1-Methylnaphthalene	221.04	0.25[e]
16	8-Methylquinoline	223.02	0.06[e]
17	1,2,3,4-Tetrahydroquinoline	225.97	0.10[c]
18	6-Methylquinoline	229.82	0.03[c]
19	1,2,2a,3,4,5-Hexahydroacenaphthylene	232.70	—[b]
20	Biphenyl	233.96	0.24[d]
21	2-Ethylnaphthalene	236.08	0.16[c]
22	1-Ethylnaphthalene	236.59	0.14[e]
23	3-Methylindole	236.66	—[b]
24	2-Methylindole	237.42	0.21[f]
25	2,6-Dimethylnaphthalene	237.58	0.17[f]
26	2,7-Dimethylnaphthalene	237.71	0.07[d]
27	5-Ethylbenzo(*b*)-thiophene	238.46	0.37[d]
28	2-Methylbiphenyl	238.77	0.04[c]
29	1,3-Dimethylnaphthalene	240.25	0.16[e]
30	1,4-Naphthoquinone	240.82	0.04[c]
31	1,7-Dimethylnaphthalene	240.66	0.25[c]
32	1,6-Dimethylnaphthalene	240.72	0.09[d]
33	2,2′-Dimethylbiphenyl	241.94	—[b]
34	2,6-Dimethylquinoline	242.43	0.29[e]
35	2,3-Dimethylnaphthalene	243.55	0.19[c]
36	1,4-Dimethylnaphthalene	243.57	0.16[f]
37	1,5-Dimethylnaphthalene	244.98	0.16[c]
38	Diphenylmethane	243.35	0.11[f]
39	Acenaphthylene	244.63	0.19[g]
40	2,2′-Bipyridyl	245.48	0.27[c]
41	1,2-Dimethylnaphthalene	246.49	0.30[e]
42	1,8-Dimethylnaphthalene	249.52	—[b]
43	2-Ethylbiphenyl	250.85	—[b]
44	Acenaphthene	251.29	0.14[g]
45	4-Methylbiphenyl	254.71	0.17[d]
46	3-Methylbiphenyl	254.81	0.15[f]
47	2,3-Dimethylindole	255.48	—
48	Dibenzofuran	257.17	0.05[c]
49	2-Methyl-1,4-naphthoquinone	259.23	0.11[c]
50	2,3,6-Trimethylnaphthalene	263.31	0.12[c]
51	1-Methylacenaphthylene	265.24	0.02[d]
52	2,3,5-Trimethylnaphthalene	265.90	0.14[e]
53	Dibenzo-p-dioxin	267.27	0.20[h]
54	Fluorene	268.17	0.15[g]

Table 39 (Continued)
PAH RETENTION INDICES

Compound no.	Compound name	Average index	Standard deviation
55	Trans-1,2,3,4,4a,9a-hexahydrodibenzothiophene	269.67	0.37[f]
56	Cis-1,2,3,4,4a,9a-Hexahydrodibenzothiophene	271.39	0.27[f]
57	3,3′-Dimethylbiphenyl	271.87	—[b]
58	9-Methylfluorene	272.38	0.17[c]
59	2,3,5-Trimethylindole	272.57	—[b]
60	4,4′-Dimethylbiphenyl	274.59	—[b]
61	5H-Indeno(*1,2-b*)pyridine	279.31	0.19[c]
62	Xanthene	280.48	0.22[c]
63	9,10-Dihydroanthracene	284.89	0.19[e]
64	9-Ethylfluorene	284.99	0.20[c]
65	9,10-Dihydrophenanthrene	287.09	0.16[c]
66	1,2,3,4,5,6,7,8-Octahydroanthracene	287.69	0.20[c]
67	2-Methylfluorene	288.21	0.15[d]
68	1-Methylfluorene	289.03	0.04[d]
69	1,2,3,4,5,6,7,8-Octahydrophenanthrene	292.03	0.06[c]
70	1,2,3,4-Tetrahydrodibenzothiophene	294.30	0.12[c]
71	9-Fluorenone	294.79	0.26[d]
72	Dibenzothiophene	295.81	0.03[d]
73	1,2,3,4-Tetrahydrophenanthrene	297.21	—[b]
74	Phenanthrene	300.00	
75	Anthracene	301.69	0.08[d]
76	Benzo(*h*)quinoline	302.22	0.11[d]
77	9,10-Dihydroacridine	304.33	—[b]
78	Acridine	304.50	0.05[c]
79	1,2,3,4-Tetrahydrocarbazole	306.76	0.19[d]
80	Phenanthridine	308.79	0.23[a]
81	Benzo(*f*)quinoline	309.25	0.19[d]
82	Carbazole	312.13	—[b]
83	9-Ethylcarbazole	313.97	0.13[d]
84	1-Phenylnaphthalene	315.19	0.05[c]
85	1,2,3,10b-Tetrahydrofluoranthene	316.37	0.14[d]
86	9-n-Propylfluorene	318.01	0.20[c]
87	3-Methylphenanthrene	319.46	0.12[e]
88	2-Methylphenanthrene	320.17	0.12[d]
89	3-Methylbenzo(*f*)quinoline	320.77	0.24[d]
90	2-Methylanthracene	321.57	0.12[c]
91	o-Terphenyl	321.99	0.16[d]
92	4H-Cyclopenta(*def*)-phenanthrene	322.08	0.15[d]
93	9-Methylphenanthrene	323.06	0.24[d]
94	4-Methylphenanthrene	323.17	—[b]
95	1-Methylanthracene	323.33	—[b]
96	1-Methylphenanthrene	323.90	0.08[c]
97	2-Methylacridine	324.46	0.05[c]
98	9-n-Butylfluorene	328.99	0.37[d]
99	9-Methylanthracene	329.13	0.19[d]
100	4,5,9,10-Tetrahydropyrene	329.69	0.24[c]
101	4,5-Dihydropyrene	330.01	0.02[d]
102	Thianthrene	330.13	0.26[e]

Table 39 (Continued)
PAH RETENTION INDICES

Compound no.	Compound name	Average index	Standard deviation
103	Anthrone	330.53	0.67[c]
104	2-Phenylnaphthalene	332.59	0.14[e]
105	9-Ethylphenanthrene	337.05	0.08[d]
106	2-Ethylphenanthrene	337.50	—[b]
107	3,6-Dimethylphenanthrene	337.83	0.14[d]
108	2,7-Dimethylphenanthrene	339.23	0.16[d]
109	1,2,3,6,7,8-Hexahydropyrene	339.38	0.17[c]
110	6-Phenylquinoline	342.45	0.04[c]
111	Fluoranthene	344.01	0.16[g]
112	9-Isopropylphenanthrene	345.78	0.28[c]
113	1,8-Dimethylphenanthrene	346.26	0.27[c]
114	2-Phenylindone	347.47	0.07[c]
115	Indeno(*1,2,3-ij*)isoquinoline	347.57	0.19[c]
116	9-n-Hexylfluorene	348.54	0.10[c]
117	9-n-Propylphenanthrene	350.30	0.17[c]
118	Pyrene	351.22	0.08[d]
119	9,10-Dimethylanthracene	355.49	0.03[c]
120	Benzo(*lmn*)phenanthridine	358.53	0.28[c]
121	9-Methyl-10-ethylphenanthrene	359.91	0.19[c]
122	m-Terphenyl	360.73	0.03[c]
123	Benzo(*kl*)xanthrene	361.38	0.15[c]
124	4H-Benzo(*def*)carbazole	364.22	0.12[d]
125	p-Terphenyl	366.10	0.16[g]
126	Benzo(*a*)fluorene	366.74	0.13[g]
127	11-Methylbenzo(*a*)fluorene	367.04	0.14[d]
128	9,10-Diethylphenanthrene	367.97	0.15[d]
129	1-Methyl-7-isopropylphenanthrene	368.67	0.19[c]
130	Benzo(*b*)fluorene	369.39	0.15[c]
131	4-Methylpyrene	369.54	0.13[c]
132	2-Methylpyrene	370.15	0.44[d]
133	4,5,6-Trihydrobenz(*de*)-anthracene	370.86	—[b]
134	1-Methylpyrene	373.55	0.11[c]
135	3,5-Diphenylpyridine	373.79	0.18[e]
136	5,12-Dihydronaphthacene	381.56	—[b]
137	9,10-Dimethyl-3-ethyl-phenanthrene	381.85	0.18[e]
138	9-Phenylcarbazole	382.09	—[b]
139	1-Ethylpyrene	385.35	—[b]
140	2,7-Dimethylpyrene	386.34	0.06[d]
141	1,2,3,4,5,6,7,8,9,10,11,12-Dodecahydrotriphenylene	386.36	0.29[e]
142	11-Benzo(*a*)fluorenone	386.41	0.06[d]
143	1,1′-Binaphthyl	388.38	0.23[d]
144	Benzo(*b*)naphtho(*2,1-d*)thiophene	389.26	0.16[e]
145	Benzo(*ghi*)fluoranthene	389.60	0.06[d]
146	Benz(*c*)phenanthrene	391.39	0.40[e]
147	Benz(*c*)acridine	392.50	0.13[e]
148	9-Phenylanthracene	396.38	—[b]
149	Cyclopenta(*cd*)pyrene	396.54	0.08[d]
150	Benz(*a*)anthracene	398.50	0.08[c]
151	Benz(*a*)acridine	398.74	—[b]
152	Chrysene	400.00	

Table 39 (Continued)
PAH RETENTION INDICES

Compound no.	Compound name	Average index	Standard deviation
153	Triphenylene	400.00	0.01[c]
154	Benzo(*a*)carbazole	401.81	—[b]
155	1,2′-Binaphthyl	405.38	0.05[c]
156	7-Benz(*de*)anthrene	406.54	0.34[d]
157	9-Phenylphenanthrene	406.90	0.14[c]
158	Naphthacene	408.30	0.22[d]
159	Benzo(*b*)carbazole	410.12	—[b]
160	11-Methylbenz(*a*)anthracene	412.72	—[b]
161	2-Methylbenz(*a*)anthracene	413.78	0.44[c]
162	1-Methylbenz(*a*)anthracene	414.37	0.17[c]
163	1-n-Butylpyrene	414.87	0.12[d]
164	1-Methyltriphenylene	416.32	—[b]
165	9-Methylbenz(*a*)anthracene	416.50	0.20[d]
166	3-Methylbenz(*a*)anthracene	416.32	—[b]
167	9-Methyl-10-phenyl-phenanthrene	417.16	—[b]
168	8-Methylbenz(*a*)anthracene	417.56	0.04[c]
169	6-Methylbenz(*a*)anthracene	417.57	0.30[c]
170	3-Methylchrysene	418.10	0.17[d]
171	5-Methylbenz(*a*)anthracene	418.72	—[b]
172	2-Methylchrysene	418.80	—[b]
173	12-Methylbenz(*a*)anthracene	419.39	0.45[d]
174	4-Methylbenz(*a*)anthracene	419.67	—[b]
175	5-Methylchrysene	419.68	—[b]
176	6-Methylchrysene	420.61	0.04[c]
177	4-Methylchrysene	420.83	0.16[c]
178	2,2′-Biquinoline	421.12	—[b]
179	1-Phenylphenanthrene	421.66	0.31[c]
180	1-Methylchrysene	422.87	0.06[c]
181	7-Methylbenz(*a*)anthracene	423.14	—[b]
182	o-Quaterphenyl	423.63	—[a]
183	2,2′-Binaphthyl	423.91	—[b]
184	2,(2′-Naphthyl)-benzo(*b*)thiophene	428.11	0.37[d]
185	1,3-Dimethyltriphenylene	432.32	—[b]
186	1,12-Dimethylbenz(*a*)-anthracene	436.82	—[b]
187	Benzo(*j*)fluoranthene	440.92	—[b]
188	Benzo(*b*)fluoranthene	441.74	0.48[c]
189	Benzo(*k*)fluoranthene	422.56	—[b]
190	7,12-Dimethylbenz(*a*)-anthracene	443.38	0.12[c]
191	1,6,11-Trimethyltriphenylene	446.24	—[b]
192	Dinaphtho(1,2-b; 1′,2′-d)-furan	450.20	—[b]
193	Benzo(*e*)pyrene	450.73	0.17[c]
194	Dibenzo(*c,kl*)xanthene	451.57	—[b]
195	Benzo(*a*)pyrene	453.44	—[b]
196	Perylene	456.22	0.29[c]
197	1,3,6,11-Tetramethyltriphenylene	461.72	—[b]
198	3-Methylcholanthrene	468.44	—[b]
199	m-Quaterphenyl	472.81	0.23[c]
200	Indeno(*1,2,3-cd*)pyrene	481.87	0.09[c]
201	Pentacene	486.81	—[b]
202	p-Quaterphenyl	488.18	0.38[d]

Table 39 (Continued)
PAH RETENTION INDICES

Compound no.	Compound name	Average index	Standard deviation
203	Dibenz(*a,c*)anthracene	495.01	0.08[b]
204	Dibenz(*a,h*)anthracene	495.45	—[a]
205	Benzo(*b*)chrysene	497.66	—[a]
206	Picene	500.00	
207	Benzo(*ghi*)perylene	501.32	0.18[e]
208	Dibenzo(*def,mno*)chrysene	503.89	—[b]
209	2,3-Dihydrodibenzo(*def,mno*) chrysene	503.91	—[b]

[a] 1 determination. [b] 2 determinations. [c] 3 determinations. [d] 4 determinations. [e] 5 determinations. [f] 6 determinations.[g]

Reprinted with permission from Lee, M. L., Vassilaros, D. L., White, C. M., and Novotny, M., *Anal. Chem.*, 51, 768, 1979. Copyright by the American Chemical Society.

observed in this work. Chrysene and triphenylene were not resolved and the benzofluoranthenes as well as many of the methylated PAH were only partially resolved.

The capillary column gc parameters for analyses of environmental samples for PAH as reported by Meuser et al.[178] at the recent symposium on PAH were as follows: 15 m × 0.28 mm glass capillary coated with SE-52, oven temperature, 100 to 320°C at a rate of 3°C/min, carrier gas, hydrogen at 50 cm/sec. Although no chromatograms were yet available for publication, it was reported at the symposium that coronene could be detected within 45 min. Investigations to apply these techniques to environmental samples are currently in progress.

Lao and Thomas[189] have not been able to report any significant improvements in separation efficiency towards PAH using WCOT glass capillary columns with Dexsil® 300 and SE-52. Once again, the benzofluoranthenes and chrysene and triphenylene were unresolved.

Blomberg et al.[190-192] have attempted to solve the problem of poor stationary phase stability by synthesizing chemically bonded phenyl polysiloxanes in situ in the glass capillary column. The resulting stability of these stationary phases was explained in terms of a slight degree of cross-linking in the stationary phase and chemical bonding of the polysiloxane to the glass surface. With these columns slightly improved separation of PAH including the partial separation of chrysene from triphenylene were reported and good thermal stability (low column bleed) of the columns up to 310°C was noted. The lifetime stability of soda-lime glass capillary columns of this type were reported to be approximately 3 months when performing 10 analyses per day.

At the present, it appears that the various methyl and phenyl silicone stationary phases lack the selectivity towards PAH necessary to resolve the various isomers. The development of more highly efficient capillary columns does not obviate the search for more selective stationary phases. Perhaps greater emphasis should be placed upon the application of liquid crystal stationary phases, which are known to possess remarkable selectivity towards isomeric PAH, to capillary columns. No reports of such applications can be found in the literature.

Greater progress has been made towards the fabrication of capillary columns with enhanced thermal stability at temperatures up to 350°C. The introduction of fused silica capillary columns has led to the successful analysis of the EPA base-neutral ex-

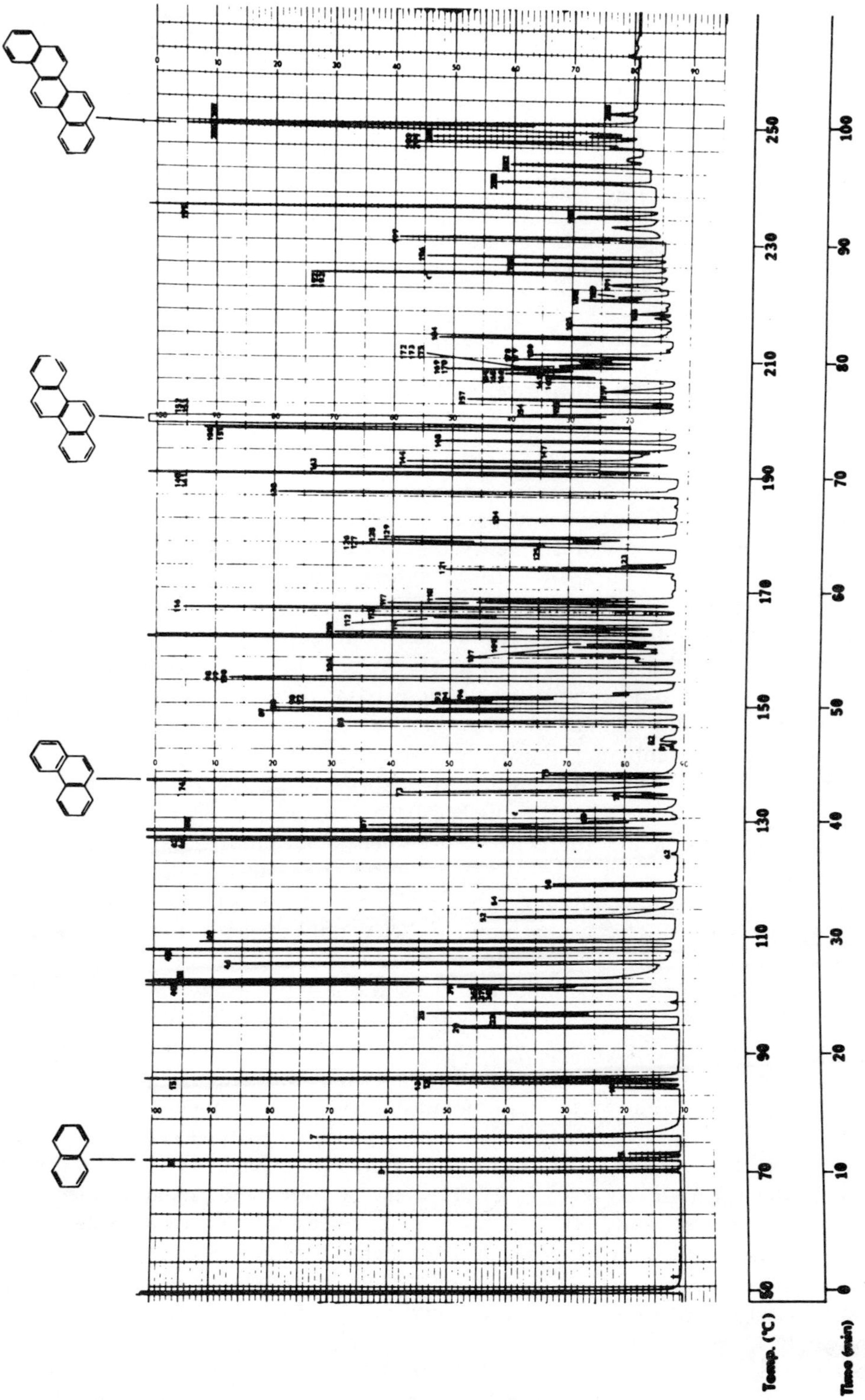

FIGURE 64. Capillary column gas chromatogram of standard PAH on a 12 m × 0.29 mm, i.d., glass capillary coated with a 0.34 μm film of SE-52. Numbered peaks refer to compounds listed in Table 39. (Reproduced with permission from Lee, M. L., Vassilaros, D. L., White, C. M., and Novotny, M., *Anal. Chem.*, 51, 768, 1979. Copyright by the American Chemical Society.)

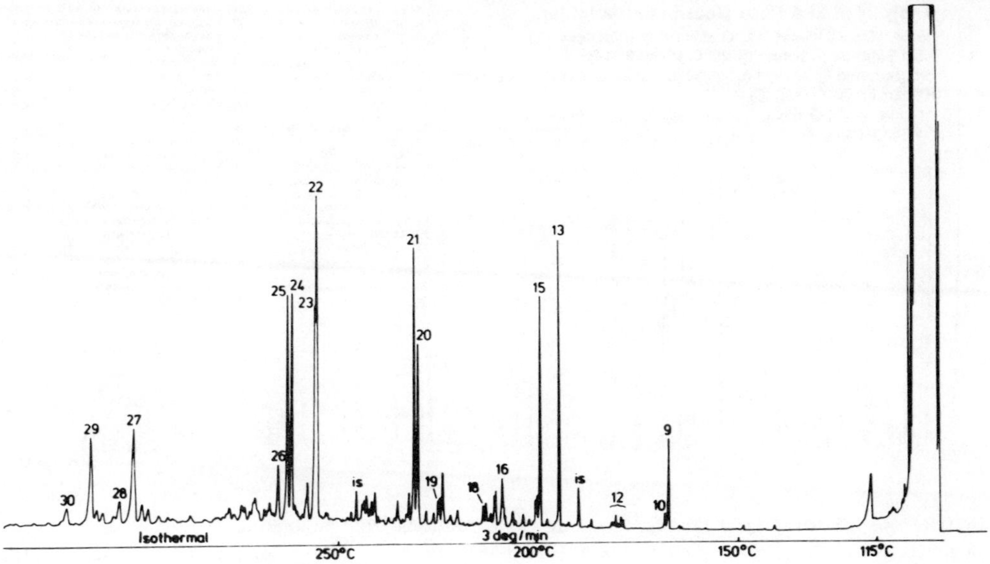

FIGURE 65. Glass capillary gas chromatogram of PAH in marine sediments outside an electrochemical plant. The stationary phase is SE-54. Numbers above the peaks refer to compounds listed in the figure key.

Key to Figure 65

9. Phenanthrene	22. Benzo (*b*) fluoranthene
10. Anthracene	23. Benzo (*k*) fluoranthene
12. Methylphenanthrenes & anthracenes	24. Benzo (*e*) pyrene
13. Fluoranthene	25. Benzo (*a*) pyrene
15. Pyrene	26. Perylene
16. Benzo (*a*) fluorene	27. Indeno (*1,2,3-cd*) pyrene
18. 1-methylpyrene	28. Dibenz (*a, h*) anthracene
19. Benzo (*c*) phenanthrene	29. Benzo (*ghi*) perylene
20. Benz (*a*) anthracene	30. Anthanthrene
21. Chrysene/triphenylene	

(From Bjorseth, A., and Eklund, G., *Anal. Chem. Acta*, 105, 119, 1979. With permission.)

tractables using a maximum column temperature of 350°C.[193] Figure 66 illustrates the low bleed level of the SE-54 capillary column at 350°C. Recently, Maskarinec and Olerich[194] reported on preliminary studies which indicate that Dexsil® 400 coated glass capillary columns are thermally stable up to 350°C. They indicated that these capillaries should be useful for analyses of high molecular weight PAH.

D. Methods of Detection

The capillary or packed column used for the separation must be coupled to a suitable detector. Four detector systems which have been applied to the determination of PAH are: (1) flame ionization, (2) gas phase fluorescence, (3) photoionization, and (4) mass spectrometry. Since gc/ms analysis is a separate technique, it will be discussed in a separate section.

The flame ionization detector, FID, responds with high sensitivity to essentially all organics and is the most commonly used detector. The design and characteristics of a flame ionization detector are discussed by Grob.[195]

For the determination of PAH, two characteristics of the FID are important: (1) sensitivity to low levels of PAH, and (2) selectivity towards PAH. In practice, the FID meets the first requirement. The second cannot be achieved since the FID responds to all organics and not just PAH.

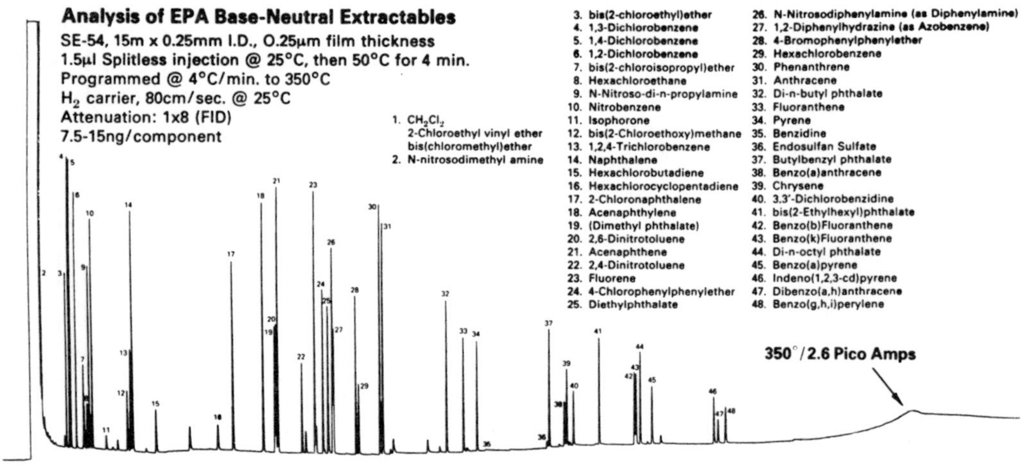

FIGURE 66. Illustration of the bleed level at 350°C resulting from an analysis of EPA Base-Neutral Extractables using a fused silica glass capillary. (From Manufacturer's chromatogram, J & W Scientific, Inc., Orangevale, Calif. 1980. With permission.)

Quantitative analytical results from flame ionization detection depend on analysis of the chromatographic peaks and for this purpose electronic integration is most commonly employed [35,116] because it is more accurate and precise than other methods.[195] For the quantitation of the chromatographic peaks, the area of the peak is compared to the peak area of an internal standard,[116,196] and 1-chlorotetradecane,[116] and octacosane[196] are commonly used internal standards for this purpose. Quantitative PAH results are obtained by using a linear relationship between peak areas (PAH/internal standard) and the concentration ratio (PAH sample/internal standard).

Limits of detection are defined in many different ways by various authors, thus they can only be used as an order of magnitude figure to compare detector capabilities. The treatment of limits of detection for hplc by Christensen and May,[159] is also valid for gc. Acheson et al.[35] were able to detect 10 ng amounts of most PAH including, B(a)P and chrysene with a FID. Basu and Saxena[34] defined detection limits as a minimum output response of five times the background noise level and a maximum sample size of 5 $\mu\ell$. They were able to detect as little as 11.9 ng of B(a)P, 10.1 ng of the benzofluoranthenes, and 14.7 ng of indeno[*1,2,3-cd*]pyrene. Burchill et al.[196] found the effective limit of detection for B(a)P to be 1.6 ng with a maximum of 1.0 $\mu\ell$ of sample. This corresponds to a peak height of 5 mm or a peak area of 100 mm at a 5 mm/min chart drive and at the lowest useable attenuation. Radecki et al.[197] reported 30 ng to be the minimum detectable amount of B(a)P using a FID. In our own studies of detection limits for PAH with a FID, detection limit was defined as a minimum peak height of two times the noise level. We were able to detect 0.1 to 10 ng amounts for individual PAH from naphthalene to coronene. As a general rule, 0.1 to 30 ng amounts of PAH are detectable with a FID.

Although fluorescence detectors are commonly used with hplc systems for determinations of PAH, they have not been widely used with gc. The main reason is that the gc effluent is a gas whereas in hplc it is a liquid. However, some studies using fluorescence detection with a gas chromatograph have been carried out.

Freed and Faulkner[198] coupled a modified Aminco-Bowman® spectrophotofluorimeter to a gc by connecting the gc to a heated fused silica flow cell (of 1.5 mℓ volume) with a heated transfer line. Fixed slits were placed into the cell holder to select various

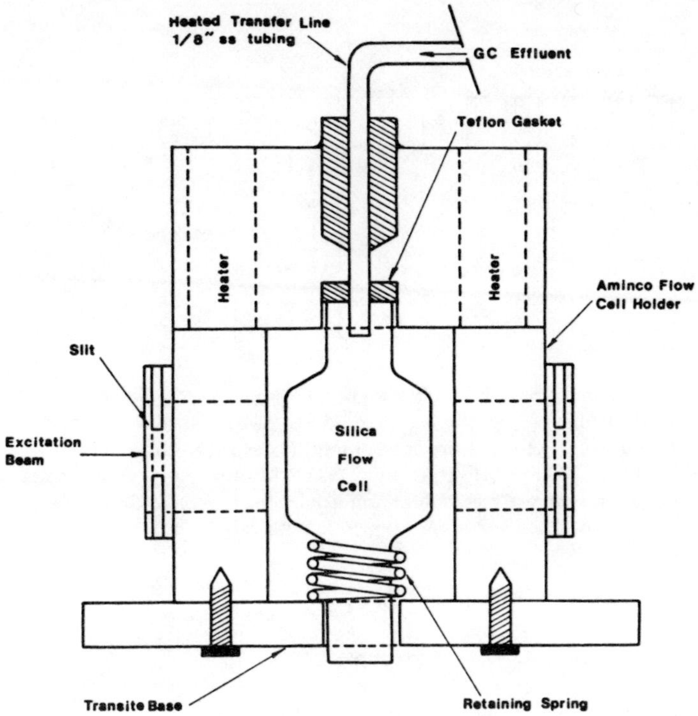

FIGURE 67. Schematic diagram of the heated fluorescence cell, as viewed from the emission monochromator. For clarity, the facing slit and the thermal insulation have not been shown. (Reprinted with permission from Freed, D. J. and Faulkner, L. R., *Anal. Chem.*, 44, 1194, 1972. Copyright by the American Chemical Society.)

bandpasses. The cell configuration is shown in Figure 67. The spectrophotofluorometer was modified to enable the spectrum from 200 to 800 nm to be scanned repetitively at 5-sec intervals and an internal standard was used to minimize light source variations on detector response. Calibration curves were linear over a range from 1 ng to 10 μg of PAH. At an excitation wavelength of 313 nm and an emission wavelength of 400 nm, 0.5 ng of pyrene was detectable when the detection limit was defined as a signal to noise ratio of two. Further instrumental details are found in the reference.

Robinson and Goodbread[199] devised a simple type of fluorescence detector for gc analyses of PAH. With their method, the sample was excited over a broad wavelength range rather than a selected wavelength, and the resulting emission was measured over a wide bandwidth. In this manner, a gain in sensitivity was achieved at the cost of reduced selectivity. The cell system and instrumental schematic are shown in Figure 68. A cylindrical cell was used in an attempt to minimize the effect of stray light. Filters were used to limit the excitation (250 to 380 nm) and emission (400 to 500 nm) wavelengths. The 380 to 400 nm range served as a neutral zone to isolate interfering scattered radiation from the Xe-Hg light source. In this experiment the detector was not actually coupled to a gc, but rather simulated conditions were used in which the PAH sample was vaporized and introduced into the cell along with carrier gas. Stopped flow and moving stream procedures were investigated and it was found that the sensitivity was greater with the former technique. By correlating peak heights and sample concentrations a value of 1.0 ng of anthracene was considered as a minimum detectable limit.

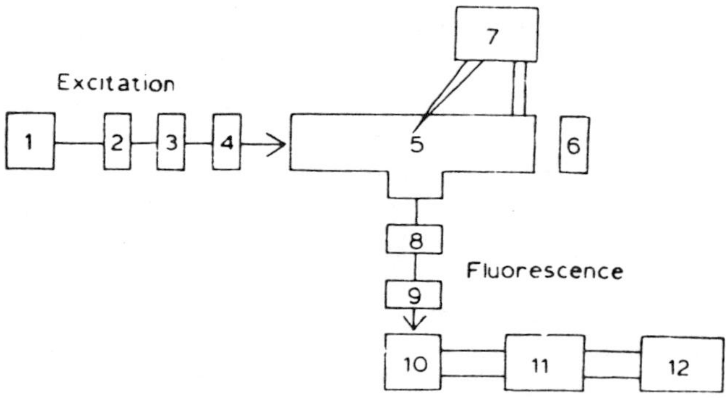

FIGURE 68. Schematic diagram of the equipment. (1) Xe-Hg lamp; (2) H_2O filter; (3) excitation filter, 250 to 380 nm; (4) lens; (5) cell; (6) mirror; (7) temperature control unit; (8) lens; (9) fluorescence filter, 400 to 500 nm; (10) P.M. tube; (11) P.M. tube power supply and amplifier; (12) X-Y recorder. (From Robinson, J. W. and Goodbread, J. P., *Anal. Chim. Acta,* 66, 239, 1973. With permission).

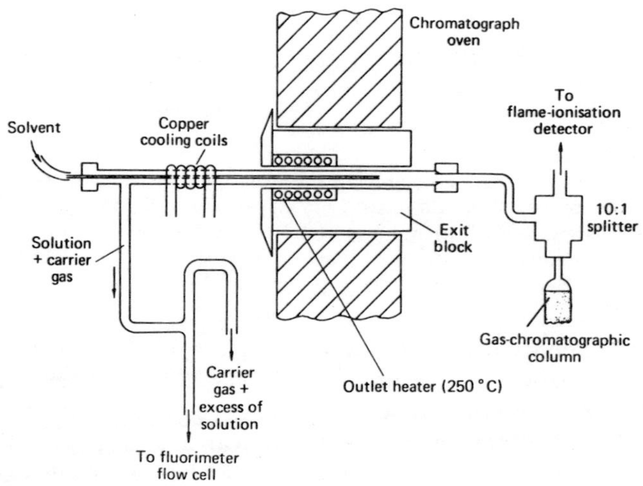

FIGURE 69. Gas chromatograph — fluorimeter interface. (From Caddy, D. E. and Meek, D. M., *Proc. Anal. Div. Chem. Soc.,* 13, 45, 1976. With permission.)

The results are comparable with those obtained by Freed and Faulker and the major advantage is that the apparatus for the wide band fluorescence method is somewhat less expensive.

Caddy and Meek[200] have also devised a simple fluorescence type of gc detector to be used simultaneously with the FID. A sample splitter was used and the portion that was directed to the fluorescence flow cell passed first to a refrigerated zone or cooler where solvent was pumped from an external source (see Figure 69). In the cooler, some solvent and solute combined and the resulting solution along with carrier gas flowed through a tube to the fluorescence cell. Along the way, some solution and carrier gas were vented by means of the T-joint shown in Figure 69. Wide band excitation (280 to 330 nm) and emission (350 to 450 nm) wavelength ranges were used. The simultaneous FID and fluorescence signals were displayed together. Peak area evaluation indicated linear response over the concentration range examined with a limit of detection of 10 ng for most PAH. Oils and sewage sludge samples were analyzed by this method.

Cooney and Winefordner[201] have definitively analyzed the instrumental effects on detection limits for gas phase fluorescence detection. They examined in critical detail the sources and optical systems which affect the signal to noise ratio. The origins of limiting noise were examined and the major contribution was considered to be stray visible light at the emission wavelength which was scattered by the sample cell and subsequently measured by the detector. They suggested that detection limits on the order of tens of picograms could be achieved if some practical methods to improve system performance could be developed and implemented. They suggested the use of highly intense and stable light sources, such as a flash pumped dye laser with maximum output in the region of maximum PAH absorption, i.e., (200 to 300 nm). Further, they recommended the use of improved optical systems such as monochromators with holographically produced gratings to be used with the laser excitation. Implementation of these recommendations would involve the optimization of grating efficiencies in the UV region and a grating design to minimize stray light.

Until the developments suggested by Cooney and Winefordner[201] are implemented, there is no sensitivity advantage in the use of fluorescence over flame ionization detection. At present, the FID is the most widely used gc detector for PAH. However, gas phase fluorescence detectors have great potential since in hplc/fluorescence systems, it has been possible to achieve detection limits for PAH in the picogram range, as compared to the nanogram range for any of the present gc detectors.

The photoionization detector, PID, was developed 3 years ago by Driscoll et al. at HNU Laboratories.[202-205] The application of a gc-PID system to analysis of alkanes, simple aromatics, PAH, and many other organics has been reported. The sensitivity of the PID system is 10 to 50 times greater than that of the FID system.[91]

In photoionization, UV radiation of sufficient energy is absorbed by a molecule to cause ionization. An 11.0 eV lamp source is sealed into the detector. Radiation from this source will ionize molecules with an ionization potential less than 11.0 eV as they elute from the column. The ions formed are accelerated to a collecting electrode. The concentration of organic molecules can be directly related to the ion current collected.

With the present radiation sources that have been incorporated into this detector compounds having ionization potentials less than about 11.0 eV can be detected by the PID method; these include aliphatic hydrocarbons (excluding methane), cyclic hydrocarbons, aromatics including PAH, and other organics. If a lamp source with an energy of 10.0 eV is used, PID detection of many solvents, e.g., methanol and chloroform can be avoided. The ability to interchange sources adds a degree of selectivity and in general a lamp with an energy output just capable of ionizing the solutes of interest is selected.

Some advantages of the PID have been listed in a pamphlet by HNU Inc.[206] These include: (1) lower limits of detection for organics by a factor of 10 to 50 over the FID, (2) a linear dynamic range of operation over several orders of magnitude of concentration, from 2 pg to 30 μg, (3) the capability of operating at a detector temperature of 300°C and a column temperature of 280°C, (4) the capability of using the detector in series with other detectors, since the PID technique is virtually nondestructive towards the sample, (5) the easy adaptability of a PID to existing gc's, and (6) the relatively low cost, approximately $3,500.

The advantages are important to analysts working with PAH. Since the modification in PID hardware to allow higher operating temperatures is relatively recent, and because HNU Inc. is the only company presently marketing PIDs, there have been very few reports in the literature on PID applications to PAH determinations. Oyler et al.[91]

have analyzed PAH and their chlorinated derivatives in water and found the PID to be 10 to 40 times more sensitive to these compounds than an FID. Detection limits of 50 to 100 pg were reported for the PAH studied: fluoranthene, anthracene, phenanthrene, and fluorene. They cited the lack of a signal for water or the acetonitrile solvent as a distinct advantage of the PID.

Driscoll et al.[207,208] have studied the relative response factors of PID and FID for lower molecular weight hydrocarbons and pesticides. In all cases, the response of the PID was greater than the FID; this results in lower limits of detection when PID was used. The greatest increase in sensitivity was observed for aromatics and pesticides. The PID was 35 times more sensitive to aromatics than the FID, and seven times more sensitive to alkanes such as n-hexane.

Hester and Meyer[209] have used the PID to detect low molecular weight aromatics in airborne particulate matter. At maximum sensitivity, with detection limits defined as an S/N ratio of two, benzene, toluene, and m-xylene could be detected at levels of less than 1 ppb.

The advantages of PID should lead to its increased application as a detector for gc. The introduction of the high temperature PID should prove especially valuable for the determination of the higher molecular weight PAH. A further advantage of the nondestructive feature of PID is that this detector can be used in series with a mass spectrometer and we predict that reports of this application will appear shortly.

E. Gas Chromatographic Detection with a Mass Spectrometer

When the mass spectrometer is used as a detector for the gc, an output similar to that from a gc-FID system is obtained. The interfacing is a problem in that the effluent from the gc must be stripped of much of the carrier gas before being introduced into the mass spectrometer. A number of detailed discussions on how this can be accomplished are available.[210,211] Once the mass spectrometer and gas chromatograph are interfaced, the mass spectrometer records the mass spectra of the gc effluent as elution from the column proceeds, and these mass spectra are stored in a data system. All the ions in each spectrum are summed and recorded as a function of elution time or mass spectrum number. The plot of the total ion current versus time or spectrum number gives a trace similar to that of an FID output. There may be some differences in intensity of response that depend on the difference in FID and mass spectrometer sensitivities to a particular compound. The retention times of the gc will not be affected if the system is properly designed.

The use of a mass spectrometer as a gc detector, though not as sensitive as some other detectors, provides a detail of data unmatched by other techniques. Figure 70 illustrates the different methods of data presentation which can be obtained with a gc/ms data system. Not only can the mass spectrum that is being scanned as well as the total ion chromatogram be presented in real time on a split screen as they are being generated, but many other types of output that are useful to the analyst can be produced. Three dimensional spectra involving mass, intensity, and time can be generated.

Because of the vast amounts of data produced by the mass spectrometer, most modern gc/ms systems are equipped with a dedicated computer to provide the output desired. The mass spectrometer scans the gc effluent at preset intervals, e.g., every 5 sec, over the course of the elution of the sample and the spectra are stored in the computer. At the end of the gc separation, several hundred spectra may be stored in the computer.

The computer can normalize peaks, subtract background column bleed, and plot the mass spectra in addition to acquiring data. One form of computer assisted data presentation is the generation of a mass chromatogram from the mass spectra stored in the data system. The intensity of a preselected ion in each spectrum stored is plotted as a function of spectrum number. These mass chromatograms are constructed from

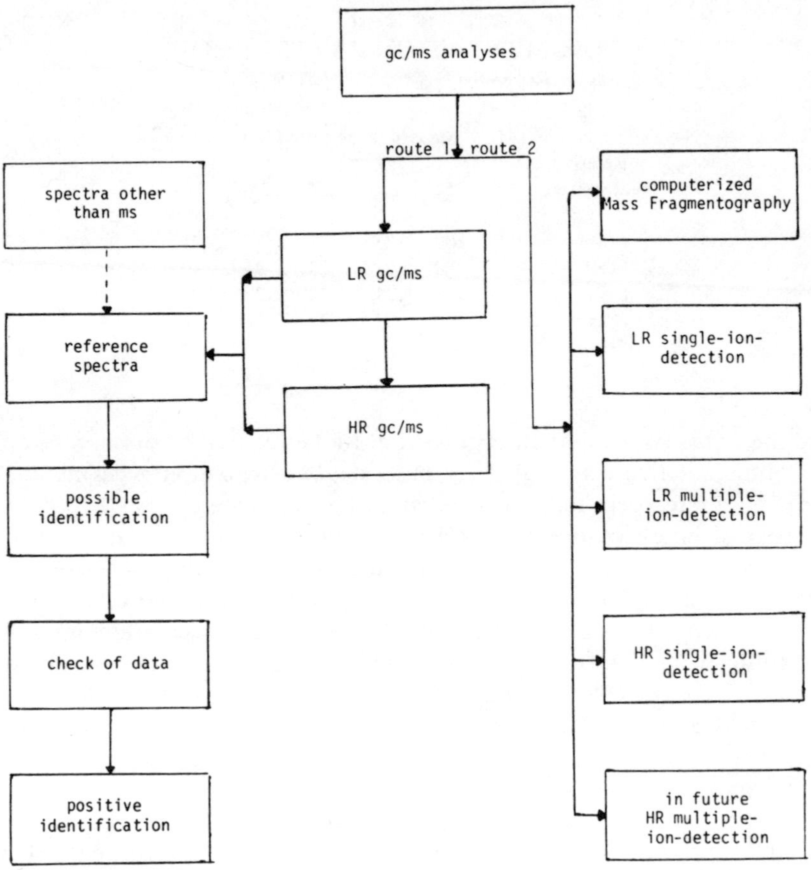

FIGURE 70. Flow chart of the different procedures applied in gc/ms analyses. Route 1, multicomponent analysis of complex mixtures; route 2, qualitative and quantitative trace analysis of complex mixtures. (From DeBrauw, M. C. T., *J. Chromatogr.*, 165, 207, 1979. With permission.)

a characteristic ion mass, e.g., a mass chromatogram of m/e:252 would serve to identify the elution time of all $C_{20}H_{12}$ PAH. The sensitivity of the mass chromatogram is less than that of the total ion chromatogram, because only a fraction of the ions of each spectrum are plotted, however, the advantage of being able to select the peaks of the chromatogram that contribute to a particular mass overcomes this drawback. In the case of PAH where the molecular ion is very intense, this loss in sensitivity is not serious.

In cases where the chemical composition(s) of the sample is(are) known in advance, another gc/ms technique, selected ion monitoring, SIM, where the intensity of a preselected characteristic ion is measured, can be used. Details on this method have been presented by Fenselau.[212] In SIM the ion intensities at a preselected m/e value are measured as the sample is eluted from the gc. By monitoring m/e:252 in a PAH analysis, the elution times of all the $C_{20}H_{12}$ PAH isomers can be determined. An increased level of sensitivity is achieved with SIM since the total scan time is dedicated to the collection of mass 252 ions rather than to measuring all the masses in the spectrum. This technique becomes multiple ion monitoring when it is expanded to include a number of characteristic ions which can be used to identify the components of a mixture.

Extensive reference libraries of mass spectra are available to assist in the interpretation of the large amount of data acquired in a gc/ms run. Several papers[213-215] have

Table 40
FRAGMENTATION PATTERNS OF SOME $C_{20}H_{12}$ ISOMERS

Relative intensities	Mass/charge ratio (m/e ratio)				
	252	126	253	250	125
B(a)P	100	23	21	16	15
B(e)P	100	21	23	25	21
Perylene	100	21	22	21	17
B(k)F	100	27	23	20	16
B(b)F	100	23	23	19	15
B(j)F	100	22	23	21	19

discussed the problems of data interpretation. McGuire et al.[215] discuss the approaches the Environmental Protection Agency have devised to simplify gc/ms analysis and subsequent data interpretation of organic pollutants in water.

Mass spectral interpretation of PAH and their derivatives is relatively straightforward and has been discussed by Borwitzky and Schomburg.[180] Unsubstituted PAH as well as the monomethyl derivatives exhibit very intense molecular ions, which are usually the base peaks. A distinction between isomeric PAH is generally quite difficult if not impossible from the examination of the fragmentation patterns. The prominent peak intensities for the PAH isomers of MW:252 are listed in Table 40. It can be seen that the identification of the isomers must be made in conjunction with elution time data.

The mass spectra of the unsubstituted PAH exhibit hydrogen cleavage of the molecular ion yielding M-1 and M-2 ions. M-26, M/2e, the doubly charged molecular ion, and (M-26)/2e ions are also observed. For PAH with CH_2 linkages, e.g., fluorene and acenaphthene, the molecular ion remains the base peak followed by M-1, M-2, M-3, M-4 as a result of hydrogen cleavage. The ions M-26 as well as M/2e and (M-26)/2e are also found.

Finnigan and Knight[216] have employed glass capillary gc/ms in analyses of Alaskan oil samples to determine the PAH content. Figure 71 shows the total ion chromatogram of the oil sample. A total of 1825 scans were taken over a period of 45 min. By searching all the recorded mass spectra for masses 178, 228, and 252, the scan number associated with PAHs of these molecular weight were identified.

The specific ion search is exemplified in Figure 72 where a total ion chromatogram overlapped with a specific ion search for m/e:178 is shown. Anthracene and phenanthrene give rise to almost identical mass spectra, therefore standards or exact retention times are needed to verify which of the two isomers is being identified.

Benoit et al.[217] have identified a number of PAH in Ottawa drinking water using gc/ms. The XAD-2 resin extract concentrate was injected onto a 3% OV-17 packed column for separation. The total ion chromatogram contained a number of weak and poorly defined peaks which made the task of obtaining identifiable mass spectra, free of interfering ions, difficult even with the aid of a computer generated background subtraction routine. To identify individual components, mass chromatograms were generated at selected ion currents by a computer search of the accumulated data for the ion of interest. The retention times for each PAH were determined by the analysis of a standard solution and the data from the XAD-2 extract was searched for the particular ion within the appropriate retention time window. The molecular ion and the next most abundant ion were selected as the characteristic ions for each PAH. A particular PAH compound was considered to be identified if the two characteristic

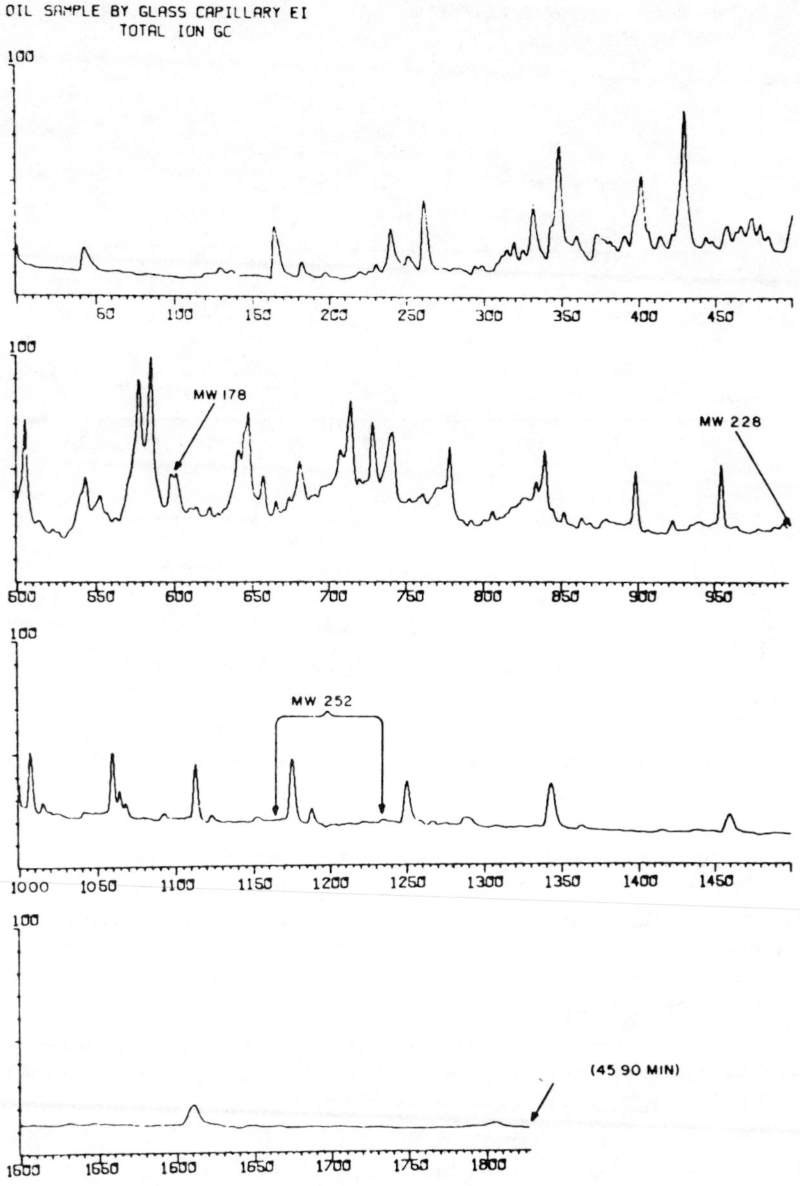

FIGURE 71. Total ion chromatogram of Alaskan oil sample. The m/e values marked on the chromatograms show where specific PAH were found by limited mass search. Conditions: 20 m × 0.02 in i.d. SE-30 glass capillary; column temperature, ambient to 250°C at 4°C/min; carrier gas, He at 2 mℓ/min flow rate; injector temperature, 250°C; sample size, 0.2 μℓ (splitless injection); mass range scanned, 40 to 350 amu; scan time, 1.227 sec. (From Finnigan, R. E. and Knight, J. B., *Identification and Analysis of Organic Pollutants in Water,* Keith, L. H., Ed., Ann Arbor Science Publishers, Ann Arbor, MI., 1975, 185. With permission.)

ions of the compound eluted from the column within the retention time window (± 0.1 min) of the reference standard and exhibited the relative abundance ratio (± 20%) observed in the pure compound. In most cases, this technique yielded positive identification. In certain instances co-eluting isomers with similar mass spectra could not be

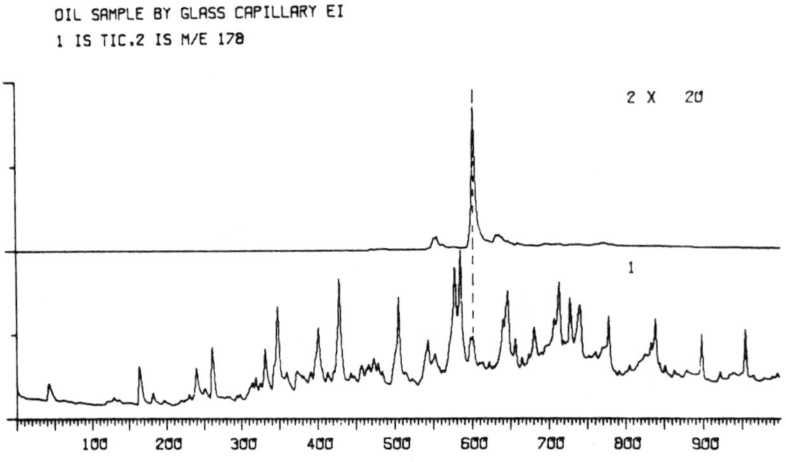

FIGURE 72. Total ion chromatogram overlapped with specific ion search for m/e = 178. (From Finnigan, R. E. and Knight, J. B., *Identification and Analysis of Organic Pollutants in Water,* Keith, L. H., Ed., Ann Arbor Science Publishers, Ann Arbor, MI., 1975, 185. With permission.)

chromatographically resolved to allow unambiguous identification. Concentrations of individual PAH which were detected ranged from 0.05 to ca. 8 ng/ℓ.

Strosher and Hodgson[168] analyzed Great Lakes water and sediments for their PAH content by gc/ms. The gc peaks which contained more than one PAH were resolved by repetitive oscillograph scanning over their respective gc elution times. For one peak, which consisted of contributions from three benzofluorene isomers, measurements were made of the relative abundances of fragment and doubly charged ions of each isomer. Rapid oscillograph scanning of the mass range from 70 to 320 m/e for the entire time span of the gc peak was sufficient to reveal the necessary differences in ionic abundance to determine the order of isomer elution. These studies, which could by means of elaborate computer methods,[168] yield quantitative measurements of multicomponent peaks were restricted to qualitative peak identifications.

Janini et al.[172] have investigated the use of BPhBT liquid crystal packed columns in gc/ms analyses. Using the SIM mode at m/e of 252, it was possible to detect a sample injection of 4 ng of B(a)P. They stated that optimization of instrumental parameters could lower this limit to less than 1 ng. When the mass spectrometer was used in the normal mass scanning mode, 30 ng of B(a)P was the detection limit.

Detection limits for PAH using column packings with lower levels of column bleed than BPhBT should be ca. 0.01 to 0.1 ng for normal mass scanning modes and ca. 1 to 10 pg for SIM modes.

The most important advantage of gc/ms over gc and hplc for analyses of PAH may be its cost-effectiveness. Finnigan et al.[218] and Budde and Eichelberger[219] have addressed this issue for the analyses of organics in the environment. Both groups concluded that gc/ms is less costly on a cost per analysis basis when a large number, ca. 100 or more, PAH or other organics are to be determined. This result was obtained by considering the capital costs of purchasing the necessary equipment to run the analyses as well as the operator costs and the overhead. Positive identification in gc cannot be made unless authentic samples are analyzed. The time and effort spent in this procedure results in a cost factor twice that of using a gc/ms system. In addition to presenting a detailed breakdown of comparison costs of gc versus gc/ms, Finnigan et al.[218] found through an industrial survey that gc/ms gave more reliable determinations of PAH than gc.

VI. SUMMARY

The type of sample which is to be analyzed and the number of compounds of analytical interest define the chromatographic approach to be used. A number of chromatographic methods have been developed for the analysis of the PAH fraction of the organic compounds isolated from a water sample. The number of PAH can range from 6 to 200 compounds. When a cleanup procedure is used, the six World Health Organization target PAH[5] can be successfully analyzed on a nanogram level by thin-layer chromatography and it would be unncessary to use hplc, gc or gc/ms to monitor just these six compounds. However, when the sample types grow in complexity and can include 20 or more PAH techniques such as hplc, gc, or gc/ms must be considered.

High performance liquid chromatography has been the most successful means of separating isomeric PAH. Unfortunately, at this time only ca. 20 PAH peaks can be resolved by hplc without overlap. The Environmental Protection Agency has listed 16 PAH on their list of Priority Pollutants[5a] and a complete resolution of these 16 compounds by hplc has been achieved by Ogan et al.[140] Picogram amounts of PAH can be determined by hplc coupled with fluorescence detection. A certain amount of cleanup is necessary to isolate the PAH from the other organics but fluorescence detection provides additional selectivity.

The complete analysis of environmental samples with as many as 200 PAH requires the use of capillary column gc. Although the establishment of a system of retention indices for capillary column gc by Lee et al.[184] is a step in the direction of total gc analysis without mass spectrometric identification, at the present time only computerized gc/ms systems are capable of analyzing complex environmental samples.

Any laboratory engaged in the analysis of environmental samples should have available thin layer chromatographic methods which can be used for both clean-up and analytical purposes.

Gc/ms analyses suffer some loss of sensitivity and selectivity towards PAH as compared to hplc but a much broader range of sample complexity can be handled by this technique and this is our preferred method of analysis. A larger initial investment is necessary for gc/ms. Cost effectiveness studies find gc/ms more economical than gc with conventional detectors.[218] A study of this type would probably also find hplc less economical than gc/ms for samples with 200 or more PAH. Development of hplc/ms systems might eventually lead to a method rivalling gc/ms in broad ranged applicability.

Chapter 5

SPECTROSCOPIC METHODS OF ANALYSIS FOR PAH

I. INTRODUCTION

Several new techniques based on spectroscopic principles are being developed for analyses of PAH. The main reason for the increasing interest in these techniques is that the separation step is eliminated, thus a one step analysis is achieved. Although it seems unlikely that a one step analysis for all of the PAH of environmental interest will ever be possible, techniques such as room temperature phosphorimetry, synchronous luminescence spectroscopy, and others will find use in partial analyses of various PAH. The new spectroscopic methods are reviewed with an emphasis on their applicability to environmental analyses.

II. ROOM TEMPERATURE PHOSPHORIMETRY

Room temperature phosphorimetry, (RTP), developed extensively by Winefordner et al.[220,221] at the University of Florida is a promising method for rapid and sensitive detection of individual PAH.[222,223]

Normally, the measurement of phosphorescence is possible only at cryogenic temperatures such as 77K, where the solute is locked into the solvent matrix to prevent collisional deactivation. However, at room temperature, the solute can achieve this same rigidity by adsorption onto the surface of various filter papers which provide a matrix-like environment analogous to that achieved at 77K. This effect, RTP, is observable with PAH.

Experimental details can be found in the References 221, 222 for an RTP analysis, which consists of four basic steps. These steps are illustrated in Figure 73. The substrate which was found to give the optimum response is filter paper, and in particular Schleicher and Schuell 604 (12.5 cm circles) and 591-C (1-in. width paper roll) were the choice of over 20 different types of paper support studied. The pretreatments of the filter paper which were studied such as moistening the paper with 1M sodium acetate, citrate, and malonate followed by drying proved to give poorer results in terms of signal to noise ratio.[221] Therefore, no pretreatment was used. As suggested in Figure 73 a continuously feeding filter paper roll could be designed to provide an automatic analysis sytem.

The sample of PAH in solution usually requires pretreatment with a heavy-atom perturber whose function is to encourage the population of the triplet state by means of intersystem crossing and spin-orbit coupling.[223] Therefore, the sample solution of PAH is adjusted with 0.2 *M* silver nitrate or thallium acetate solution. Silver and thallium were found to give the greatest enhancement of phosphorescence intensity. Once the pretreatment is complete, as little as 3 $\mu \ell$ of sample is spotted on the filter paper.

The next step involves a predrying process in which an infrared heating lamp is used. The infrared lamp was found to be more efficient than the use of desiccators, blowers, or ovens.[223] While the analysis is taking place, it is necessary to have a source of warm, dry air blowing over the sample. It has been noted that the optimum flow gas in nitrogen.[221] The measurement can be made on a spectrophotofluorometer modified for the addition of a phosphoroscopic attachment, details of which are found in Reference 220.

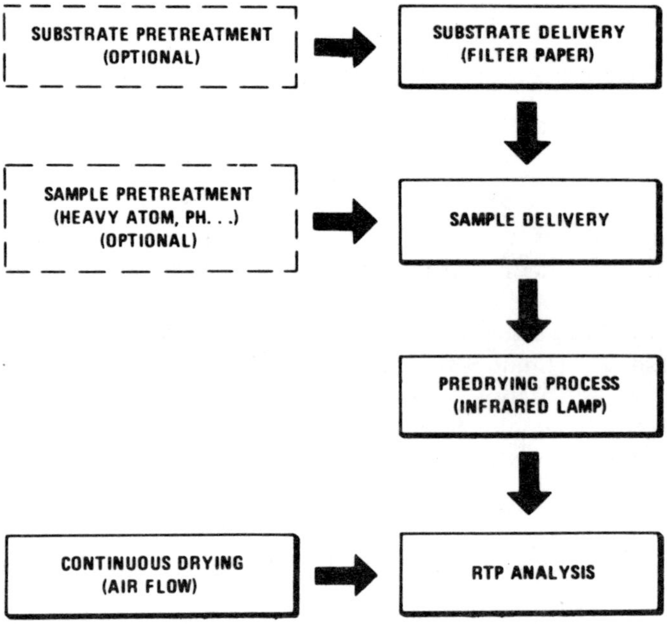

FIGURE 73. Block diagram of the RTP analytical scheme.

The limits of detection range from about 8 ng for 1, 2, 5, 6 dibenzanthracene down to 0.45 ng. of chrysene. Relative SD are less than 5%. In all cases, the PAH were studied individually and not in mixtures. Analysis time could be adjusted depending on rates of drying to allow for seven to eight measurements per minute. Further investigations are being conducted to lower the limits of detection.

Clearly, the advantage of RTP is the ability to obtain a fast, quantitative estimate of the amount of a particular PAH present in a sample. For environmental applications, this advantage is of paramount importance. Problems exist which place an inherent limit on the applicability of the method, namely (1) benzo(*a*)pyrene is not detectable due to its very short phosphorescence lifetime[220] (2) multicomponent mixtures are not capable of being analyzed unless prior separation is performed or it is established that the phosphorescence intensities of the other components do not interfere with the analyte of interest — a fact seldom realized, (3) limits of detection of PAH are only comparable with techniques such as hplc and tlc with fluorescence detection, and (4) phosphorescence instrumentation is not any less expensive than that required for a tlc analysis which can also provide PAH separations. These limitations should not discourage the use of RTP. Developments in synchronous spectroscopy will extend the technique to samples with more than one PAH and the limits of detection should improve with research on instrumental and procedural variations. However, for now, RTP is in a developmental state and application to environmental analyses of PAH awaits further improvements in the field.

III. SYNCHRONOUS LUMINESCENCE SPECTROSCOPY

In a conventional fluorescence or phosphorescence spectrometer, an emission spectrum is generated by scanning the wavelength of emission, λ_m, as the substance is excited at a single excitation wavelength, λ_x. Analogously, an excitation spectrum is formed by scanning the excitation wavelength while recording the luminescence intensity at a single emission wavelength. A synchronous spectrum is generated if both λ_m

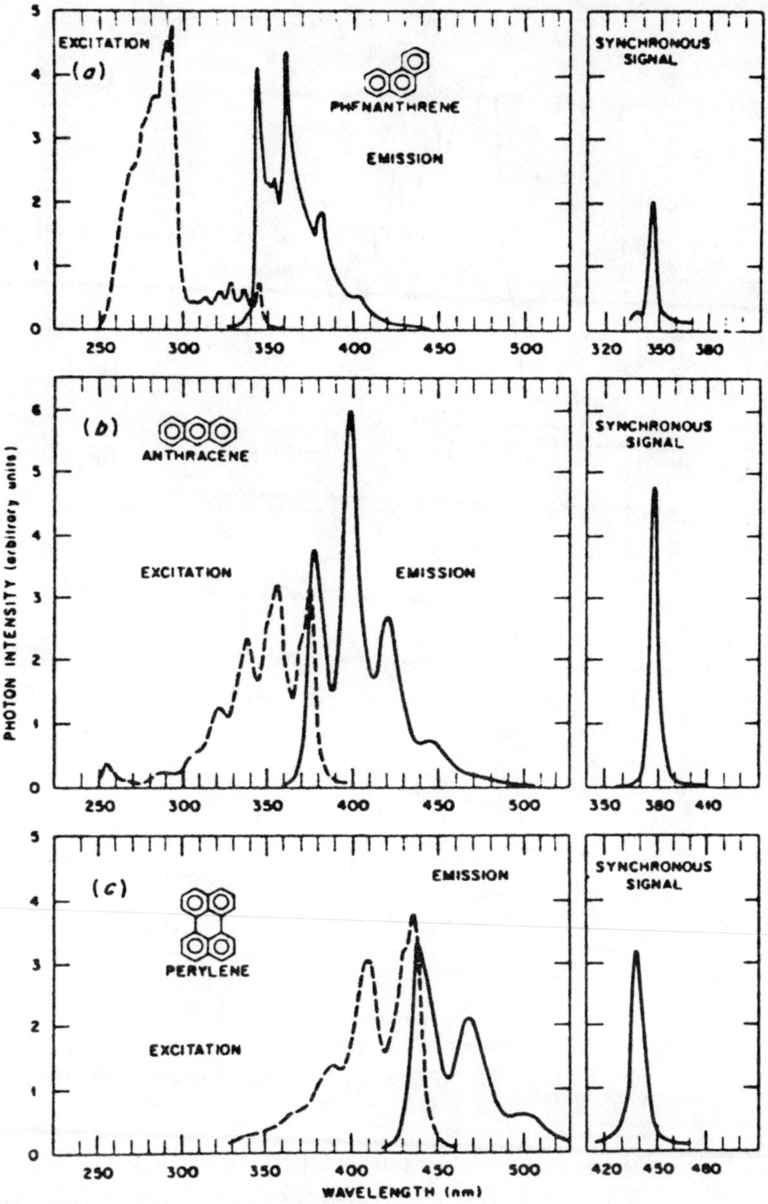

FIGURE 74. (a) Fluorescence excitation, emission, and synchronous spectra of phenanthrene, (b) Fluorescence excitation, emission, and synchronous spectra of anthracene, (c) Fluorescence excitation, emission, and synchronous spectra of perylene.

and λ_x are scanned with the wavelength interval between the two, $\Delta\lambda$, kept constant. The nature of the resulting synchronous spectrum will naturally be changed. The result is that the synchronous luminescence intensity depends on the character of the normal excitation and emission spectrum as well as the wavelength difference $\Delta\lambda$ between the excitation and emission wavelength. The intensity can be expressed as: $I_s = KE_x(\lambda_m - \Delta\lambda)E_m(\lambda_m)cd$ where E_x is the excitation spectrum at a wavelength of $(\lambda_m - \Delta\lambda)$, E_m is the normal emission spectrum, c is the analyte concentration, d the sample thickness, and K a characteristic luminescence constant. A more completely derived treatment of the above is found in References 224 and 225.

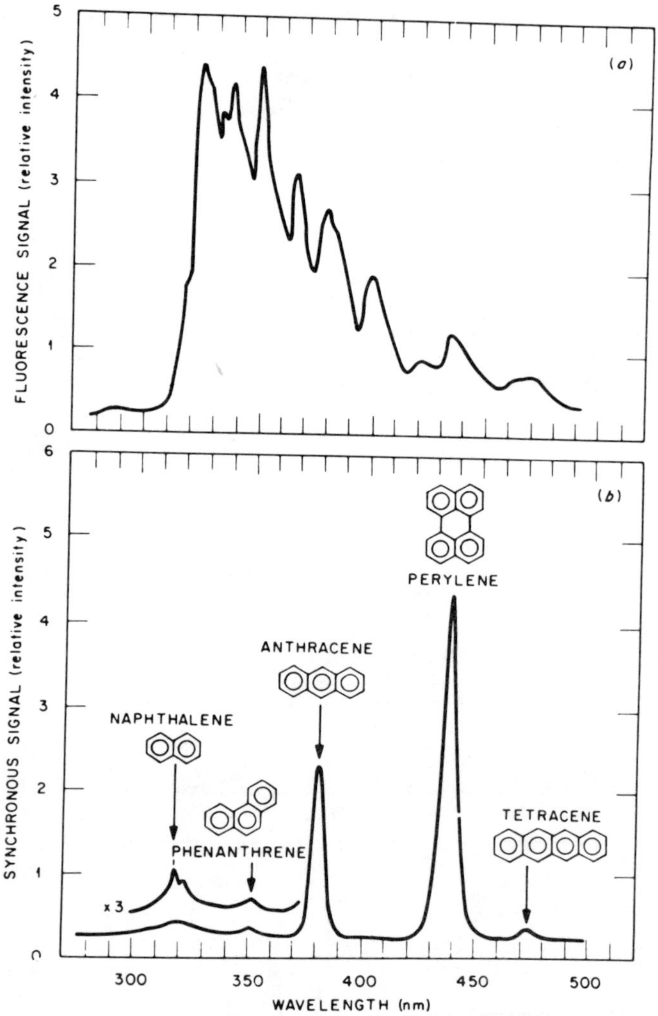

FIGURE 75. (a) Fluorescence spectrum of a mixture of naphthalene, phenanthrene, anthracene, perylene, and tetracene ($\lambda_{exc.}$ = 258 nm), (b) Synchronous signal ($\Delta\lambda$ = 3 nm) of the same mixture.

Figure 74 illustrates the difference in physical appearance between the normal excitation and emission spectra for PAH and the resulting synchronous luminescence spectra. The major difference is the simplicity of the synchronous signal — one peak resulting from the entire excitation and emission spectrum. Whereas the normal wavelength range spanning the excitation and emission curves is about 200 nm, the synchronous signal bandwidth is between 10 to 20 nm. It is the prospect of simplifying a resultant spectrum of several components into individual peaks per component that makes the technique attractive.

The $\Delta\lambda$ parameter depends on the individual characteristics of the PAH excitation and emission spectra and a full treatment of the effect of this variable on the synchronous spectrum's characteristics can be found in Reference 225. The important features for the analyst interested in PAH are the following: (1) multicomponent mixtures of PAH can be analyzed from the resulting simplified spectra, and (2) the $\Delta\lambda$ parameter can enable the "selective" monitoring of one PAH in the presence of others if $\Delta\lambda$ is unique to one PAH. An example of the spectral simplification is shown in Figure 75. The top half of the figure is the conventional fluorescence emission spectrum of a mixture of five PAH where it is quite clear that identification of individual PAH is quite difficult. The resulting synchronous spectrum shows five peaks, each of which can be correlated with an individual PAH.

The instrumental requirements for synchronous luminescence spectroscopy are no different from any other fluorescence or phosphorescence analysis. A spectrofluorometer must be used which has the capability of allowing the excitation and emission monochromators to be linked and scanned together; i.e., synchronously.

The technique has been developed for rapid screening of aqueous pollutants and has been applied to monitor naphthalene and its methyl derivatives in the by-product water of the Synthane gasifier.[223,226] 1-methyl and 2-methylnaphthalene could be monitored in the presence of one another with 1 ppb detection limits.[226]

It is possible to apply the synchronous technique to RTP studies to add another degree of selectivity. The added factor of selectivity is based upon the single-triplet energy difference for the PAH. This energy difference denoted as Δst is usually unique for a particular species. The phosphorimetric approach is exactly the same as that described earlier in the section on RTP analysis except for the conversion of the excitation and emission monochromators to the synchronous mode. Using the synchronous technique along with the unique Δst values for each PAH, several isomeric PAH can be analyzed in the presence of each other.[226,227]

The basic approach is to determine optimum Δst values for each PAH and to set the wavelength interval $\Delta\lambda$ in the synchronous approach equal to this singlet-triplet wavelength difference. In the analysis of a mixture of 1, 2, 3, 4 and 1, 2, 5, 6 dibenzanthracene,[227] (DBA), the energy differences Δst were found to be 274 and 252 nm, respectively. In a mixture of 42 ng of 1, 2, 3, 4, DBA and 4.2 ng. of 1, 2, 5, 6 DBA the synchronous RTP spectra were clearly differentiated and thus able to be quantified. This fact is important since 1, 2, 5, 6 DBA is noted to be more highly carcinogenic than its isomer. Although it is stated that the analysis of benzo(*a*)pyrene and benzo(*e*)pyrene should be easy by synchronous RTP due to a large difference in Δst values, no results are shown for such an analysis. In fact, it has been stated in a paper also published in 1978 by Bower and Winefordner that benzo(*a*)pyrene does not phosphoresce at room temperature due to a very low lifetime (τ_{77K} = 111 ms.).[221] Since the experiment conditions of Bower and Winefordner and Vo-Dinh and Gammage[227] are essentially identical it seems unlikely that B(a)P can be monitored phosphorimetrically at room temperature. However, it seems likely that many isomeric PAH may be amenable to such an analysis if Δst values are sufficiently different and phosphorescence lifetimes are favorable at room temperature.

Some applications of synchronous RTP include the analysis of pyrene in a synthoil sample.[227] Nanogram detection limits are reported.[226]

The potential of synchronous RTP seems greater than synchronous fluorescence spectroscopy due to the greater degree of selectivity afforded by the selection of a Δst value. The limitations are naturally the ability of the PAH to exhibit a favorable phosphorescence at room temperature.

In general, the synchronous spectroscopy methods are likely to find utility in qualitative PAH analyses. Although a number of attempts have been made to apply the techniques more fully to environmental analyses, it seems that its advantages at present do not make it an obvious replacement for gc and hplc methods which are as sensitive to low levels of PAH and which can obviously more easily handle multicomponent samples.

Latz, et al. have pointed out limitations in the use of synchronous luminescence spectroscopy as a quantitative technique.[228] Some of the difficulties they noted were coincidences in wavelength and/or weak fluorescence signals. They observed that the signal for fluorene could be quenched or effectively nulled by the presence of pyrene at the same concentration. Even with the concentration of pyrene, only one third of the concentration of fluorene results in a 50% depressed fluorene signal. They were concerned with the possible loss of information resulting from the spectral simplification achieved by the synchronous technique. Of concern was the effect of impurities, overlapping bands and coincidental wavelength differences, and the quenching effect mentioned. They caution investigators against the use of synchronous luminescence spectroscopy as more than a qualitative finger-printing tool and note that even using the technique in this mode requires the conventional fluorescence excitation and emission spectra in addition.

Though the warnings of Latz et al. are severe, synchronous methods are just developing and should continue to find use as routine monitors for some PAH in water samples. Further research is definitely needed before it grows in importance as a tool for the analysis of complex environmental samples of PAH.

IV. MATRIX ISOLATION FLUORESCENCE AND INFRARED ANALYSIS

Since room temperature fluorescence and phosphorescence spectra of PAH in solution are broad and relatively structureless, multicomponent analyses of PAH by these methods are difficult unless the analyst is fortunate enough to be able to apply synchronous luminescence, wavelength modulation, or derivative techniques to his studies. However, even with these aids, any realistic total analysis of mixtures of PAH likely to be found in environmental sources is essentially impossible.

Measurements of fluorescence and infrared absorption at cryogenic temperatures are easier since both the fluorescence and infrared spectra show more fine and resolved structure at temperatures of 77K and below. Since the spectral peaks are decreased in width, it is possible to analyze some multicomponent mixtures of PAH. The matrix isolation, (MI), technique is one of two methods developed for low temperature spectroscopic studies.

Matrix isolation studies have been carried out almost exclusively by a number of workers at the University of Tennessee.[229-233] The basic requirement for a solute to be studied by an MI method is that it be convertible to the gaseous phase. With temperatures ranging from 40 to 100°C, vacuum sublimation of PAH is possible.[230] The PAH vapor is then allowed to effuse through a Knudsen orifice or molecular leak where it is mixed with large excess ($10^3 - 10^8$:1 on a mole basis) of a matrix gas. Nitrogen is claimed to be a better choice than argon.[231] This gaseous mixture is then condensed on a cold surface held at 20 K by the use of a closed-cycle helium refrigerator. The solid sample condensed onto a sapphire or cesium iodide window is capable of being analyzed spectroscopically by a fluorescence or infrared technique. The main reason for this approach is to produce a sample in which the solute is homogeneously dispersed in the matrix. If the ratio of matrix to solute is high enough, then the average distance between solute molecules should be large enough to prevent the occurrence of quenching and energy transfer between solute molecules. A matrix: solute ratio of 10^4 is about average, (on a mole basis). Further experimental details can be found in the references cited (229 to 233).

MI fluorescence spectroscopy has been used to study a number of PAH, including the six isomeric monomethylchrysenes,[229,233] and hplc fractions from a Synthoil sample.[231] MI fluorescence studies were felt to be a useful diagnostic criteria for qualitative PAH identifications. Band widths for individual peaks are larger than those observed in low-temperature Shpol'skii studies discussed in the next section, but "quasilinear" fluorescence spectra essentially identical to those observed in Shpol'skii studies can be realized if organic solvents are used as matrix substances.[232]

The quantitative capabilities of MI fluorometry are similar to studies run at room temperature. In general, detection limits for PAH range from 5 to 10 pg. for B(a)P up to 50 pg for most other PAH. Analytical calibration curves exhibit a linear range of roughly three to five decades in PAH concentration. For example, the response for chrysene is linear from 20 pg to an upper limit of 5 μg, over five orders of magnitude. The relative precision of the technique using an internal standard is about 4.0%. An advantageous characteristic which is noted[231] is that the working calibration curve obtained for pure PAH samples may be directly applicable to individual PAH in a mixture sample. This would indicate that very little interference between solutes in a mixture occurs in the MI fluorescence method.

Although MI fluorescence has potential, Stroupe et al.[231] cite four limitations towards its future widespread use: (1) fluorescence of major components in a sample can effectively swamp out the response from minor components preventing their analysis, (2) impurities arising from sampling, preconcentration, and chromatographic separations preceding the measurement step can lead to unaccountable fluorescence, (3) fluorescence spectra do not lend themselves to easy structural interpretation thus standard samples for each PAH studied are necessary to allow comparisons to be made between unknown and standard, and (4) certain organic pollutants such as partially reduced PAH, heterocyclics, etc., may not give a fluorescence signal. Due to these reasons, Matrix Isolation-Fourier Transform Infrared spectroscopy, MI-FT-IR is being studied by the investigators at Tennessee as an alternative and complementary technique to MI-fluorescence. Infrared studies had not found much use in PAH studies up to this point, simply because the levels of detection were not low enough to be applicable to environmental samples. The use of FT-IR along with instrumental modifications and techniques such as beam condensing optics and ultramicrosampling have permitted detection limits as low as 50 ng for PAH.[229]

Qualitative studies of PAH by MI-FT-IR are potentially a powerful means for identifying isomeric PAH in mixtures.[230] Individual MI-FT-IR spectra of five PAH which are commonly difficult to separate chromatographically, are shown in Figure 76. Figure 77 depicts in detail the 900 to 700 cm^{-1} region for each of these PAH where the greatest differences arise and where qualitative identification can be made. The principal IR bands for the PAH studied by the MI-FT-IR method are listed in Table 41. It can be seen that the IR peak bandwidths are narrower for the MI method compared to conventional IR, and are on the order of 4 to 8 cm^{-1} maximum.

As seen from the above discussion, individual PAH spectra are readily distinguished by the MI method. The extension to a mixture analysis can be made, but is naturally more difficult since the resultant spectrum is more complicated. A five component PAH mixture analyzed by MI-FT-IR is shown in Figure 78. All of the components can be identified by at least one major absorption band in the 900 to 700 cm^{-1} region. The investigators[230] state that up to ten PAH can be identified in a mixture whereas samples of greater complexity must be chromatographically separated to some extent.

The applicability of the technique to quantitative analysis was tested by seeing if the data obeyed Beer's Law. Linearity was obtained in the range from (50 to 200 ng.) which was the limite of detection, up to a limit of about 100 μg. Detection limits of this magnitude were obtainable only with the use of an ultramicroscampling system and beam condensing optics. The use of an internal standard for quantitative studies was found to help in reducing errors resulting from experimental variations from run to run.

The MI-FT-IR technique has been applied to studies of PAH content in synthetic liquid fuels and shale oils[229] as well as to mixtures of mono-methylchrysenes.[233] As a detector for hplc or gc fractions it can be applied to many more samples provided the relative lack of sensitivity is not a serious drawback.

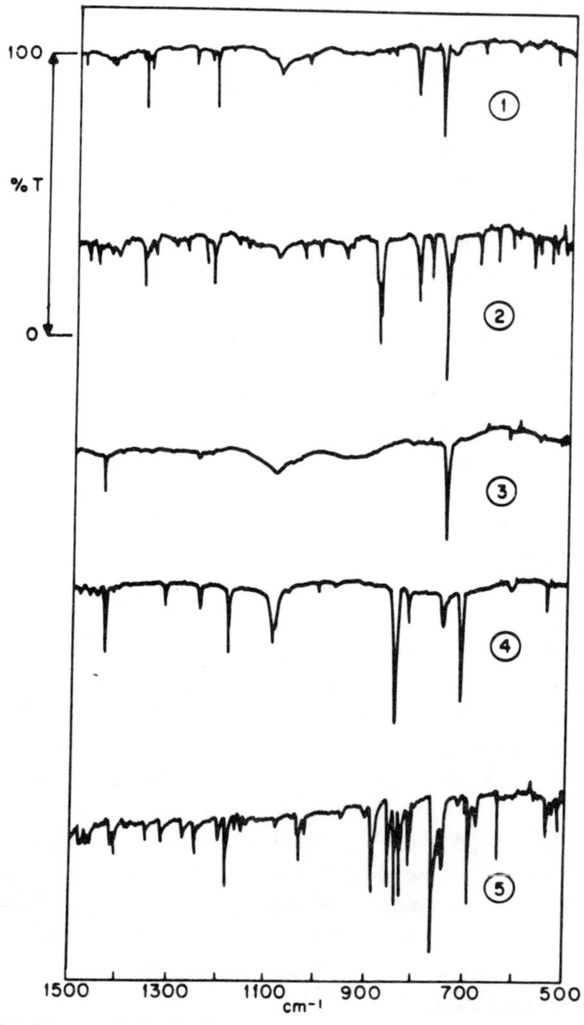

FIGURE 76. MI-FT-IR spectra of (1) chrysene; (2) benz(*a*)-anthracene; (3) triphenylene; (4) pyrene; (5) benzo(*a*)pyrene in a nitrogen matrix (nitrogen/PAH mole ratio approximately 10^3). Resolution: 2.0 cm^{-1}. Spectra are plotted vs. the vacuum reference. Note that the 1100 cm^{-1} peak is caused by an impurity on the CsI window.

Some of the future prospects outlined by Wehry et al. include the use of pulsed sample deposition in the MI technique itself, and organic solvents for MI fluorescence.[229] The utility of MI-FT-IR will naturally grow as computer techniques such as pattern recognition, applied to FT-IR, are developed more fully. The use of a computer in the technique should definitely ease the task of data handling. Any developments which extend the capabilities of fluorescence spectroscopy, e.g., laser sources and more sophisticated detectors besides photomultipliers, will also extend the usefulness of matrix isolation fluorescence techniques. The MI methods are being developed and will need further study before becoming more widely applicable to environmental samples of PAH. However, the potential of MI-FT-IR as a powerful chromatographic detection device for isomeric PAH and the use of MI-fluorescence as a "fingerprint" tool has already been realized.

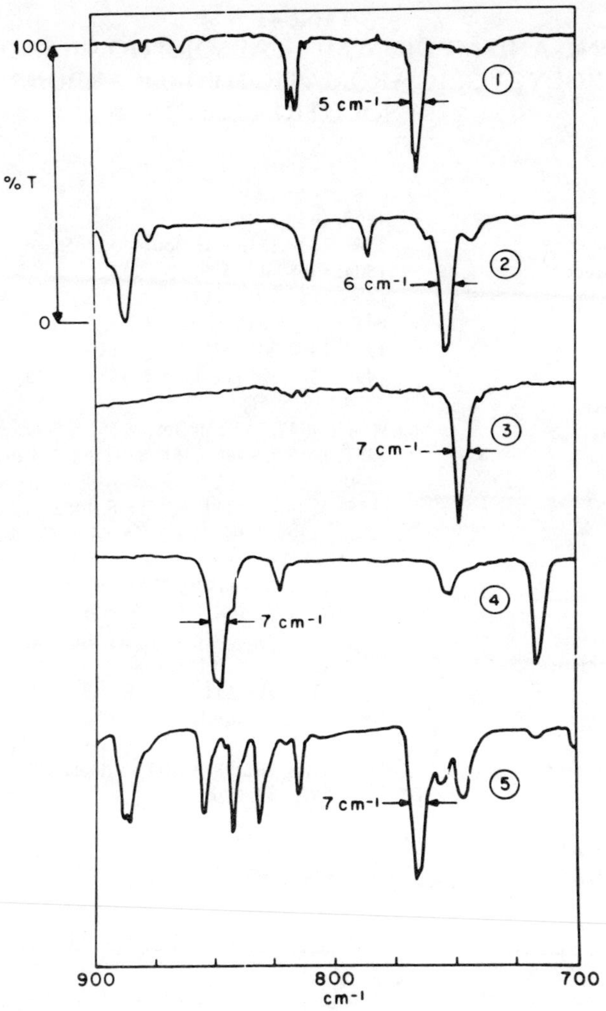

FIGURE 77. Expanded spectra in the region 900 to 700 cm^{-1} for the same compounds as in Figure 76.

V. LOW TEMPERATURE FLUORESCENCE — THE SHPOL'SKII EFFECT

Analytical studies of PAH by means of low temperature fluorescence utilizing the Shpol'skii effect have been widely published in the Russian literature of the late 1960s and early 1970s.[234] Qualitative and quantitative studies of PAH in all types of environmental, foodstuff, and model samples were carried out. Kirkbright and others[234-236] published a three part series of papers in *Analyst* reviewing most of the Russian literature pertaining to Shpol'skii effect studies on PAH. They critically examined the utility of the technique to future analytical applications. Their findings regarding the efficiency of various instrumental sample cells, light sources, and detectors are reviewed as well as their qualitative and quantitative analytical results on PAH. A number of critical remarks in the literature on the limitations of the Shpol'skii effect are also reviewed and a consensus outlook of future applications is presented.[229,231]

Table 41
STRONG AND MEDIUM MI-IR ABSORPTION BANDS OF POLYCYCLIC AROMATIC HYDROCARBONS (RESOLUTION 2 cm^{-1})

	IR Bands
Triphenylene	1502, 1438, 746[a] cm^{-1}
Chrysene	1365, 1234, (819, 816 doublet), 765[a]
Benz[*a*]anthracene	1504, 1365, 886, 809, 785, 752[a]
Pyrene	1436, 1185, 846[a], 715
Anthracene	881, (733, 730 doublet),[a] 603
Phenanthrene	1504, 1462, 870, 818, 540.5,[a] 617
Carbazole[b]	1496, 1464.5, 1454, 1397.5, 1338,[a] 1323, 1241, 754, 729.5, 618, 569
Tetracene	900.5,[a] (747, 744.5 doublet), 551.5
Fluoranthene	1461, 1455.5, 1444, 1138, 830 (782, 778 doublet),[a] 750
Fluorene	(1483.5, 1480 doublet), 1456.6, 1451.5 doublet) 1408, 1301, 1190, 1005, 958, 746[a], 696, 622
Acenaphthene	(1431, 1422 doublet), 1374, 1017.5, 843, 792,[a] 750
Perylene	1498, 1385, 818,[a] (776, 774 doublet)
Benzo[*b*]fluorene	957, 872, 775,[a] 764, 728, 570, 475
Benzo[*a*]fluorene	1223, 1020, (828, 826 doublet), 763, 760.5 doublet)[a]
1,3,5-Triphenylbenzene[b]	1580, 1558, 1503, 1417, 1080, 1032, 884, 758.5,[a] 702, 632, 623, (614, 611 doublet), 509
Benzo[*a*]pyrene	1884, (888, 886 doublet), 855, 843, 831, 765,[a] 747, 694
1,2,5,6-Dibenzanthracene	1513, 1458, 892, (813, 809 doublet),[a] (750, 745 doublet), 667, 649, 529

[a] Strongest MI-band.
[b] Not a PAH.

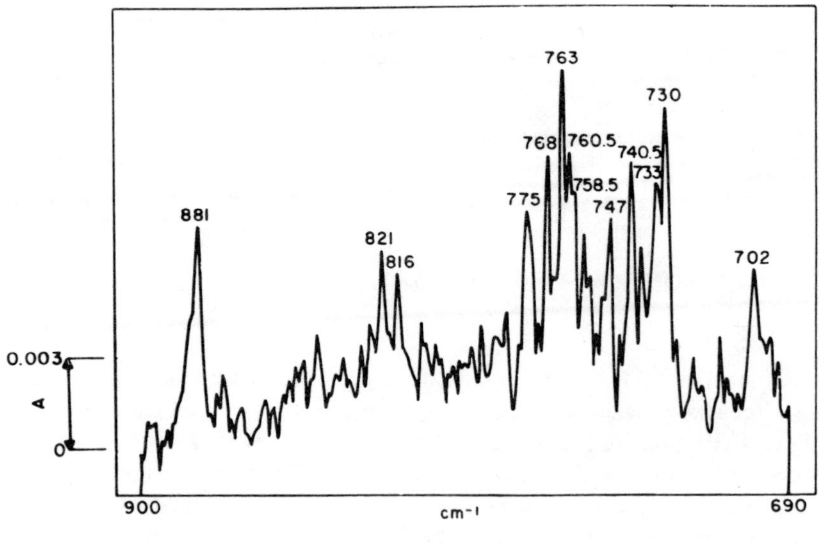

FIGURE 78. The spectrum of a five component mixture consisting of 7 μg of benzo(*a*)fluorene (753, 760.5 cm^{-1}), 7 μg of benzo(*b*)fluorene, (775 cm^{-1}), 7 μg triphenylbenzene, (758.5, 702 cm^{-1}), 10 μg anthracene (733, 730 cm^{-1}), and 10 μg phenanthrene (740.5 cm^{-1}) in nitrogen (range of nitrogen/PAH mole ratio 1.46×10^5 to 3.56×10^5). Resolution: 2.0 cm^{-1}.

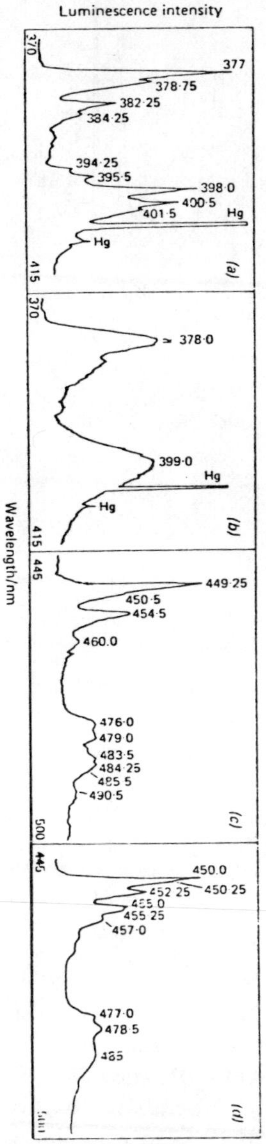

FIGURE 79. Solvent effect observed in the emission spectra at 77K for anthracene with (a) n-hexane-cyclohexane solvent and (b) n-octane — cyclohexane solvent; and for 3,4,8,9 dibenzopyrene with (c) n-octane-cyclohexane solvent, and (d) n-decane-cyclohexane solvent. (Reprinted with permission from Kirkbright, G. F. and Delima, C. G., *Analyst*, 99, 338, 1974. Copyright by the American Chemical Society.)

The Shpol'skii effect is observed if an aromatic solute, e.g., a typical PAH is dissolved in a particular n-alkane solvent, frozen at 77K, and excited at a wavelength capable of producing a luminescence spectrum. The observation that can be made and which led to analytical applications is that this luminescence spectrum exhibits an extremely well-resolved fine structure which is characteristically unique for each solute. The postulated reason for the fine structure is that in the rigid matrix, the solute molecules are incapable of interaction with each other resulting in sharp line electronic spectra.[234]

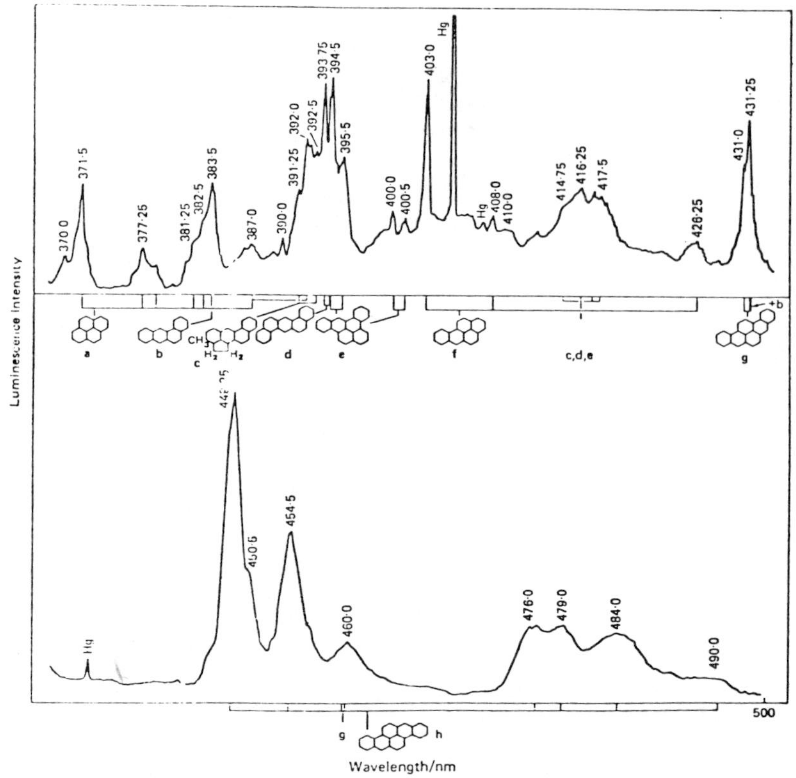

FIGURE 80. Emission spectra of a synthetic mixture in n-octane-cyclohexane at 77K; (a) pyrene; (b) 1,2 benzanthracene; (c) 3-methylcholanthrene (d) 1,2,5,6-dibenzanthracene; (e) 1,2,4,5 dibenzopyrene; (f) 3,4 benzopyrene; (g) 3,4,9,10 dibenzopyrene; and (h) 3,4,8,9 dibenzopyrene. (Reprinted with permission from Kirkbright, G. F. and Delima, C. G., *Analyst*, 99, 388, 1974. Copyright by the American Chemical Society.)

The unique feature of the effect is that the n-alkane solvent used must be particularly matched to the solute. The length of the n-alkane must be of the same molecular dimension as the solute, e.g., the PAH molecule. The degree of complexity and uniqueness of the luminescence spectrum depends upon this match. A comparison of the spectrum of anthracene in n-hexane and in n-octane is shown in Figure 79 where it can be seen that a sharp line spectrum is observed only in n-hexane. In addition, the solvent effect for 3, 4, 8, 9-dibenzopyrene is illustrated in the figure. In all cases studied, cyclohexane is used to dissolve the PAH, and the n-alkane is added to the solution such that the resultant solution contains 90% n-alkane. No effects on the observation of the sharp line spectra are caused by the presence of the cyclohexane. Solvents such as tetrahydrofuran were also studied and were observed to give rise to well-resolved fine structure luminescence spectra of PAH. A more thorough study of the solvent effect can be found in Reference 234.

The first parameter which must be decided upon is the selection of n-alkane or suitable matrix solvent. The experimental setup must then be determined. Causey et al.[235] have thoroughly addressed this issue and made a comparison study of excitation sources, sample cells, and detection systems. Xenon and medium pressure mercury vapor discharge lamps gave comparable results in terms of detection limits, and the opinion of Causey et al.[235] is that both sources should be used. However, in their studies with a helium-cadmium laser, limited to excitation at 325 and 441 nm, the

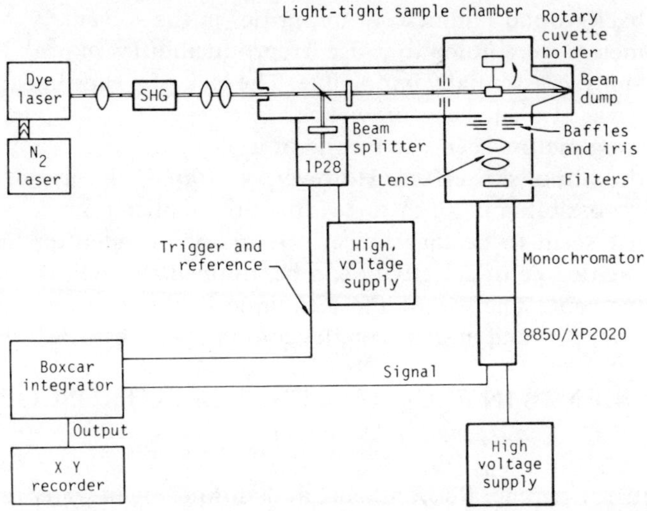

FIGURE 81. Schematic of the experimental apparatus used. (Reprinted with permission from Richardson, J. H. and Ando, M. E., *Anal. Chem.*, 49, 955, 1977. Copyright by the American Chemical Society.)

attainable limits of detection were about two orders of magnitude lower than with xenon or mercury lamps (provided the excitation wavelengths of the laser were optimum for a particular PAH). On the basis of these studies, they suggested that the use of a tunable dye laser might be very beneficial.

Effects on the luminescence spectra were shown to depend on rates of cooling and positioning of the sample cell. A Dewar-flask sampling system with a silica cell and a copper cryostat cell that are described in Reference 235 were compared. Reproducibility of peak intensities was improved with the copper cryostat cell.

D. C. integration, photon counting, and repetitive optical scanning used in conjunction with a signal averaging system were compared.[235] A gain in limits of detection was noted with the signal-integration method, but the use of this method precluded the simultaneous qualitative studies of all PAH. The advantages of photon-counting were found to be difficult to exploit. The use of repetitive optical scanning showed potential for studies at low PAH concentrations.[235]

Qualitative analyses by Shpol'skii effect low temperature luminescence spectroscopy are already a proven success. Figure 80 shows the emission spectra of a synthetic eight component mixture of PAH. It was found to be relatively easy to identify all eight components by identification of particular peaks. In fact, all of the reported literature studies indicated that as a "fingerprinting" aid the method was useful and reliable. Quantitative analyses were not as easy. Limits of detection were on the nanogram level and below for PAH but problems existed in the reproducibility and linearity of the fluorescence intensity. Overlap of solute absorption spectra giving rise to "inner filter" effects and energy transfers between the solute and other species in the sample caused inaccurate intensity values. Only through the use of a combined standard additions and internal standard approach could these errors be reduced, and even then the maximum linear response was only over a range of two orders of magnitude of concentration. Other critical comments on utilization of Shpol'skii effect spectra for quantitative analyses have appeared.[229,231] Some factors which prevent reliable quantitative analysis include formation of microcrystalline solute aggregates, different rates of freezing, fluctuations in the final temperature of the solution, differences in fluorophore concentration, and background luminescent impurities in the solvent.[231] All of the above effects and parameters were noted to cause irreproducibilities in peak intensities making quantitative analysis essentially impossible. The lack of a wide linear working range of concentrations was also cited as a drawback.

The overwhelming picture seems to be one of a bleak future as regards quantitative exploitation of the Shpol'skii effect. However, as stated, the use of the spectra as "fingerprints" is possible and since linewidths are smaller than in matrix isolation spectra they would seem to be the choice in terms of one-step qualitative analyses. Furthermore, some degree of selectivity can be attained through the use of a unique n-alkane matrix solvent. The use of the technique in this capacity should grow for polluted aqueous samples and marine samples contaminated by nearby oil spills.

VI. DEVELOPMENTS IN FLUORESCENCE SPECTROSCOPY APPLIED TO PAH

Detection limits for pyrene, fluoranthene, and anthracene in water ranging from 0.5 ppt (parts per trillion) to 4.4 ppt have been reported by Richardson and Ando.[237] Their analytical technique which involved using laser-induced molecular fluorescence should prove to be extremely useful if modified as a chromatographic detector system for PAH. With some improvements, the method could possibly be used to directly monitor PAH in ground water.

Figure 81 is a schematic of the experimental system used by the authors.[237] The novel approach involved the use of a Molectron® UV 1000 N_2 laser and a Molectron® dye laser operated in the DL 200 configuration. For anthracene, fluoranthene, and pyrene the following excitation wavelengths and dyes were employed: anthracene (258.7 nm, coumarin 500), fluoranthene (287.0 nm, rhodamine 6G), and pyrene (273.0 nm., coumarin 495). For anthracene, fluorescence excitation was also performed at 254.0 nm using lithium formate monohydrate (Lasermetrics) to double the fundamental of 508 nm. (coumarin 500).[237] Other instrumental design features included a signal averaging unit, PAR® Model 162/163/164 dual channel boxcar. Further details on the other instrumental features can be found in Reference 237.

The PAH solutions were prepared by making up stock solutions in methanol or methanol/H_2O followed by serial dilution. Glassware was investigated as a source of fluorescent impurities. The cuvettes were mounted rigidly in a reproducibly locatable rotary turret which allowed for rapid comparisons with blank solutions.

Figure 82 illustrates the fluorescence intensity versus concentration curve for pyrene, where the error bars represent the standard deviations for each of the dilutions. The only limit to linearity results from self-absorption or the point of maximum PAH solubility in water.[237] In Figure 83, it is clear for naphthalene that the linearity of response spans almost seven orders of magnitude in concentration. According to Richardson and Ando, consideration of the size of the laser beam leads to the conclusion that a minimum detection level of 0.5 ppt, such as that for pyrene, corresponds to approximately 10^7 molecules actually detected. In general it was found that the limiting factor on minimum detectability was the reproducibility in the preparation and placement of the sample.

Some research topics which were suggested to improve these detection limits include (1) better sample handling and dilution procedures, (2) consideration of oxygen quenching on the fluorescence signal, (3) improvement in sample optics and light collection, and (4) further progress in UV laser technology.

The advantage of having a direct analytical technique for PAH in water samples without prior methods of preconcentration would indicate that research will continue on laser-induced fluorescence detection, though high cost, lack of wide wavelength range, and instrumental complexities limit its present utility.

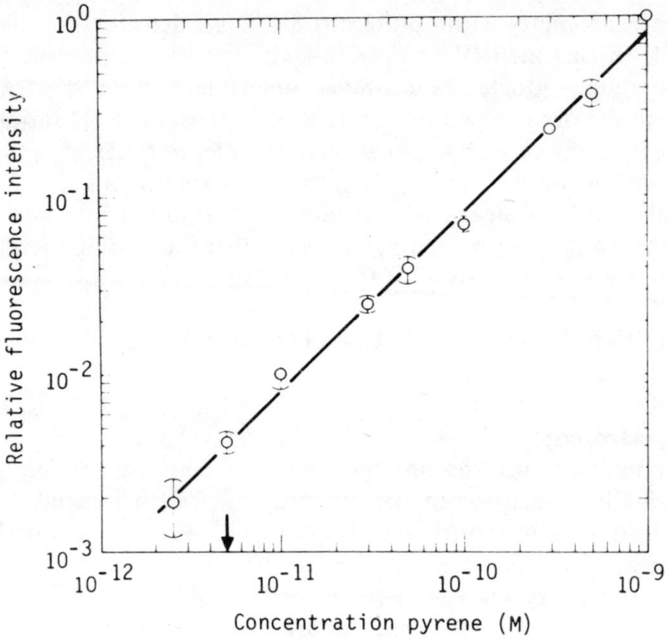

FIGURE 82. Pyrene fluorescence intensity vs. concentration. Arrow represents PPT. (Reprinted with permission from Richardson, J. H. and Ando, M. E., *Anal. Chem.*, 49, 955, 1977. Copyright by the American Chemical Society.)

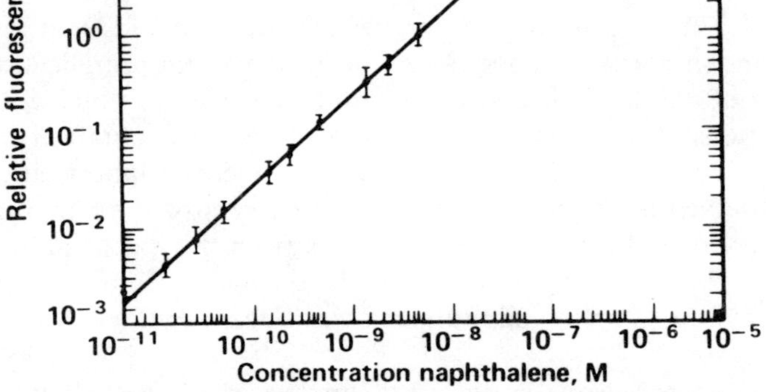

FIGURE 83. Naphthalene fluorescence intensity vs. concentration. (Reprinted with permission from Richardson, J. H. and Ando, M. E., *Anal. Chem.*, 49, 955, 1977. Copyright by the American Chemical Society.)

Fluorescence spectrometry with optoelectronic image detectors is a technique that is being developed. In this method of detection the traditional photomultiplier tube is replaced with parallel optoelectronic image detectors. Entire spectra are recorded simultaneously on the order of milliseconds with increases in S/N ratios claimed. The technique has been reviewed with application to samples of PAH.[238]

Rapid scanning fluorescence spectroscopy that is based on a video fluorometer has also been investigated as a means of providing multicomponent analysis. Johnson et al.[239] review the technique noting comparisons to other multicomponent systems. They also consider applications to analysis of PAH.

VII. DERIVATIVE AND WAVELENGTH MODULATION TECHNIQUES IN SPECTROSCOPY

A. Derivative Spectroscopy

Derivative methods of spectroscopy have been successfully applied to PAH analyses.[240-243] In derivative spectroscopy, one records the first or second derivative of luminescence intensity or absorbance with respect to wavelength versus wavelength. By means of this procedure, the detectability of minor spectral features such as weak shoulders or interfering bands can be enhanced. Some of the general experimental procedures for generating derivative spectra are reviewed and reported applications to PAH are presented.

O'Haver[240] has summarized the experimental approaches employed to obtain derivative spectra. In the case of a digitally recorded spectrum, the differentiation of the signal can be performed numerically. Otherwise, for real time application the experimentally available techniques for generating derivative spectra fall into two basic classes:[240] those systems which operate on the output signal of the spectrometer (electronic differentiation, mechanical tachometers) and those systems which operate on the light beam in the optical network of the instrument (dual wavelength spectrometry, wavelength modulation). The details of the methods are found in the references noted. The general consensus of workers in the field is that electronic differentiation offers simplicity and low cost but suffers from the inability to obtain derivatives at a fixed wavelength.[241] A decrease in the S/N ratio is also noticeable with electronic differentiation. Wavelength modulation techniques are claimed to provide a better S/N ratio.[241] A commercially available second derivative UV absorption spectrophotometer is described in Reference 243. Dual wavelength spectrophotometers capable of generating first derivative spectra are available from Perkin-Elmer Corp. and American Instrument Co.

Gammage et al.[223] have published a number of industrial and environmental monitoring applications with a second derivative spectrometer. One application is based on the measurement of naphthalene and alkylnaphthalene vapors which arise from many fossil fuel operations. A complete separation of the peaks is obtained with a mixture containing 90 ppb of naphthalene and 73 ppb. of 2-methylnaphthalene. This separation could be observed in the second-derivative spectrum of the mixture even though the two peaks were only 3 nm apart. The analysis from the spectrum indicated a concentration of 86 ± 2 ppb. of naphthalene and 71 ± 2 ppb of 2-methylnaphthalene and the analytical curve for 2-methylnaphthalene was linear from less than 40 ppb up to 240 ppb.

The same authors have also investigated the presence of 2-methylnaphthalene in the by-product water of oil shale retorting and Synthane coal gasification with a second derivative spectrometer.

Hawthorne and Thorngate[243] describe a second derivative UV absorption spectro-

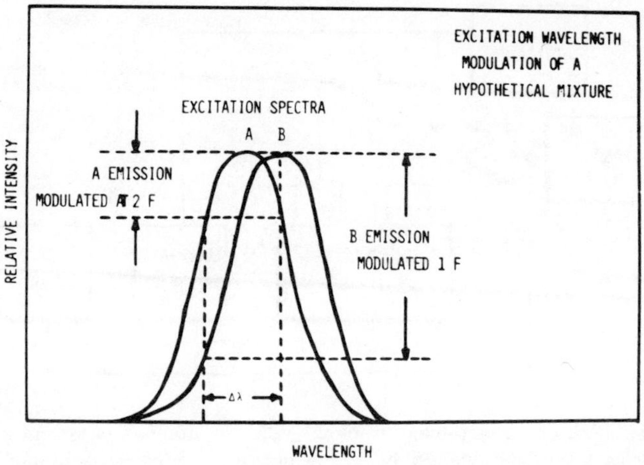

FIGURE 84. The Principle of Selective Modulation. A and B are the excitation spectra of two hypothetical molecules in a mixture. Modulation of the excitation wavelength over the indicated interval Δ λ, at a modulation frequency of F Hertz, will modulate the fluorescence intensity of B at the same frequency, while the fluorescence intensity of A will be modulated at twice this frequency (i.e., 2F). (Reprinted with permission from O'Haver, T. C. and Parks, W. M., *Anal. Chem.*, 46, 1886, 1974. Copyright by the American Chemical Society.)

photometer capable of analyzing vapor phase concentrations of the volatile PAH. Their instrument is also claimed to be useful as a monitor for aqueous pollutants.

O'Haver[240] describes a number of applications of derivative spectroscopy to PAH. In a mixture of 1.0 ppm anthracene, 21 ppb chrysene, and 9 ppb carbazole, the normal fluorescence emission band exhibits only a weak shoulder contribution from the chrysene. However, with the use of second derivative spectroscopy, as little as 6 ppb of chrysene can be detected and measured. Weak shoulders can be translated into easily quantified peaks.

The analysis of pyrene in the presence of anthracene was also discussed and the utility of second derivative spectroscopy for such an analysis was demonstrated.[240]

When applicable, derivative spectroscopy can lead to greater resolution of peaks and to the quantification of weak shoulders buried in much larger peaks. Some disadvantages of the technique include a reduction in the signal to noise ratio upon differentiation and the expense of modifying an available instrument to perform the differentiation. However, as an added tool for a UV or fluorescence chromatographic detector, derivative techniques should definitely find utility by providing another degree of selectivity in the separation of mixtures of PAH.

B. Wavelength Modulation Techniques

Figure 84 is an illustration of the basic principle behind wavelength modulation techniques developed by O'Haver and others at the University of Maryland.[240,241,244] The overlapping excitation spectra of two hypothetical substances A and B are shown in Figure 84. An individual fluorescence analysis of A or B is impossible because of the overlap of the spectra. Selective excitation which involves excitation at a wavelength unique to only one solute is not practical. In the selective modulation technique, the wavelength of the excitation monochromator is allowed to periodically modulate over the indicated wavelength interval, $\Delta\lambda$. Due to this modulation, the emission intensities are also modulated and with $\Delta\lambda$ selected in the manner shown, the emission of com-

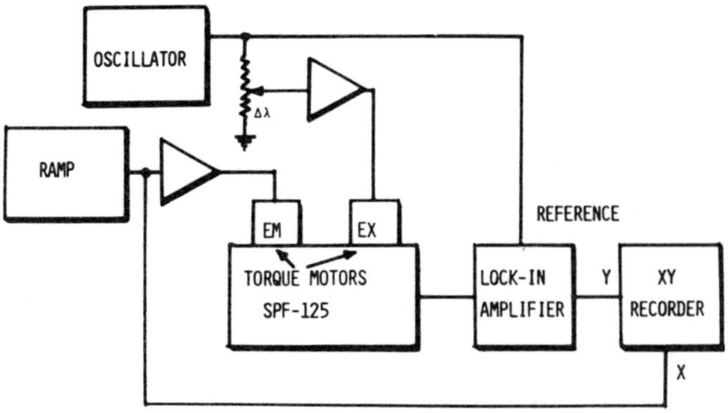

FIGURE 85. Block diagram of the selective modulation system in the excitation modulation mode. For operation in the emission modulation mode, the control lines to the torque motors are interchanged. (Reprinted with permission from O'Haver, T. C. and Parks, W. M., *Anal. Chem.*, 46, 1886, 1974. Copyright by the American Chemical Society).

ponent B is modulated at the wavelength modulation frequency, 1F, while the emission due to component A will naturally be modulated at 2F. If the detection system includes a frequency selective or lock-in amplifier, the emission of either A or B can be selectively monitored independently. Therefore, the emission spectrum of either A or B can be measured without interference from the other component. The complete principles can be found in the original reference.[244]

In general, there are two different modes of operation. There is a "null mode" in which the spectrum of one component in a binary mixture is nulled or erased. The other is an enhancement mode which can be used to selectively increase the intensity of one substance relative to others and is applicable to multicomponent mixtures. However, mixtures of more than three components are difficult to handle.

Selection and positioning of the Δλ parameter depend on the particular PAH involved. Details can be found in the references, but in general, the critical requirement which must be met is that the two excitation spectra not have exactly coincidental maxima and minima at all points. Modulation conditions can be predicted in advance but are usually determined by running a standard sample to optimize the parameters.

Figure 85 is a schematic of the instrumental apparatus required. A normal fluorescence spectrometer can be modified to allow modulation techniques to be carried out. Conventional scanning and modulation techniques are accomplished by application of the appropriate ramp and AC waveform to the motors.[240] Complete details of the apparatus can be found in References 240 and 244.

The techniques have been successfully applied to mixtures of PAH. O'Haver et al.[240,241,244] have successfully analyzed benz(*a*)anthracene in the presence of an excess of chrysene. Normally, these two PAH are difficult to resolve by conventional chromatographic techniques. As noted earlier, Fox and Staley[143] were able to determine the same two PAH in an hplc fraction with quantitative success. Various other model PAH mixtures analyzed in this manner include (pyrene-anthracene), (benz(*a*)anthracene-chrysene-triphenylene), and (chrysene-anthracene). The emission spectrum of pure chrysene excited at an excitation maximum is shown in Figure 86A[241] Upon addition of a large excess of anthracene, the resultant emission spectrum, (Figure 86B), is obtained by selective excitation at a wavelength yielding the greatest chrysene-anthracene intensity ratio. Still, it can be seen that the effect of the anthracene is still significant. However, the null-mode selective modulation designed to strip out the an-

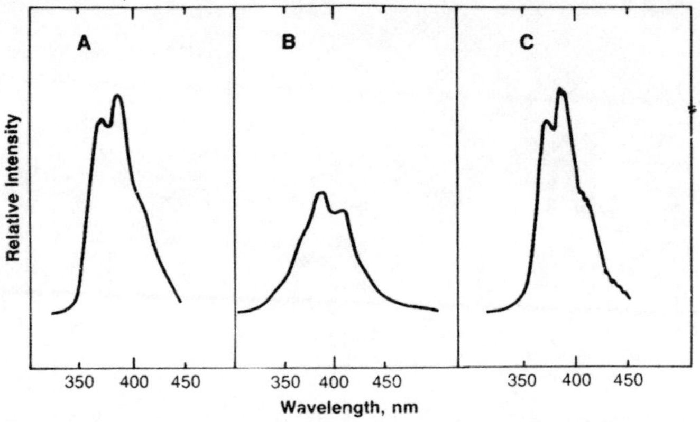

FIGURE 86. Excitation wavelength modulation to allow measurement of fluorescence emission spectrum of one component in a mixture of two fluorescent compounds whose excitation and emission spectra are extensively overlapped. (A) Emission spectrum of pure chrysene solution excited at its excitation maximum, (B) emission spectrum of a mixture of chrysene in the presence of an excess of anthracene excited near an excitation minimum of anthracene, (C) emission spectrum of same mixture obtained by excitation wavelength modulation with conditions adjusted to strip out spectrum of anthracene. (Reprinted with permission from O'Haver, T. C., *Anal. Chem.*, 51, 91A, 1979. Copyright by the American Chemical Society.)

thracene spectrum is successful in restoring the original emission spectrum of chrysene, (Figure 86C). The technique cannot result in any S/N ratio gain and the intensity of 86C is obtained only by increasing the instrumental sensitivity.

One of the apparent difficulties is this reduction in the S/N ratio. In Figure 86 the contribution of the noise is obvious. In addition, selective modulation does not alter in any way fluorescence quenching, inner filter effects, and variables which affect the slope of the analytical curve.[240] However, as a technique which requires only a moderate amount of instrumental modification, provides real-time results without the aid of a computer, and is proven to effect separations of certain isomeric PAH, selective modulation will find use in future applications.

Chapter 6

REFERENCES

1. Rose, F. J. C., *Proc. Anal. Div. Chem. Soc.*, 15, 148, 1978.
2. Gelboin, H. V. and Ts'O, P.O.P., Eds., *Polynuclear Aromatic Hydrocarbons and Cancer, Volume 1: Environment, Chemistry and Metabolism,* Academic Press, New York, 1978.
3. Jones, P. W. and Freudenthal, R., Eds., *Polynuclear Aromatic Hydrocarbons — Chemistry, Metabolism and Carcinogenesis,* Raven Press, New York, 1976.
4. Jones, P. W. and Freudenthal, R., Eds., *Polynuclear Aromatic Hydrocarbons — Chemistry, Metabolism and Carcinogenesis,* Raven Press, New York, 1978.
5. World Health Organization, *International Standards for Drinking Water,* 3rd ed., Geneva, 1971.

5a. Keith, L. H. and Telliard, W. A., *Environ. Sci. Technol.*, 13, 416, 1979.

6. Andelman, J. B. and Suess, M. J. *Bull. W.H.O.*, 43, 479, 1970.
7. Andelman, J. B. and Snodgrass, J. E., *Crit. Rev. Environ. Control,* 4, 69, 1974.
8. Hutzinger, O., Safe, S., and Zander, M., *Analabs Res. Notes,* 13, 1, 1973.
9. Harrison, R. M., Perry, R., and Wellings, R. A., *Water Res.*, 9, 331, 1975.
10. Fishman, M. J. and Erdmann, D. E., *Anal. Chem.*, 51, 317R, 1979.
11. Chian, E. S. K. and DeWalle, F. B., *J. Water Pollut. Control Fed.*, 50, 1026, 1978.
12. Hildebrand, J. and Scott, R. L., *The Solubility of Non-Electrolytes,* Dover Publications, New York, 1964.
13. Bohon, R. L. and Claussen, W. F., *J. Am. Chem. Soc.*, 73, 1571, 1951.
14. McAuliffe, C., *J. Phys. Chem.*, 70, 1267, 1966.
15. Klevens, H. B., *J. Phys. Colloid Chem.*, 54, 283, 1950.
16. Tsonopoulos, C. and Prausnitz, J. M., *Ind. Eng. Chem. Fundam,* 10, 593, 1971.
17. MacKay, D. and Shiu, W. Y., *J. Chem. Eng. Data,* 22, 399, 1977.
18. Wauchope, R. D. and Getzen, F. W., *J. Chem. Eng. Data,* 17, 38, 1972.
19. Schwarz, F. P. and Wasik, S. P., *J. Chem. Eng. Data,* 22, 270, 1977.
20. Schwarz, F. P., *J. Chem. Eng. Data,* 22, 273, 1977.
21. Aquan-Yuen, M., MacKay, D., and Shiu, W. Y., *J. Chem. Eng. Data,* 24, 30, 1979.
22. Andrews, L. J. and Keefer, R. M., *J. Am. Chem. Soc.*, 71, 3644, 1949.
23. May, W. E., Wasik, S. P., and Freeman, D. H., *Anal. Chem.*, 50, 171, 1978.
24. May, W. E., Wasik, S. P., and Freeman, D. M., *Anal. Chem.*, 50, 997, 1978.
25. May, W. E., *Diss. Abstr. Int. B,* 39, 774, 1978.
26. Davis, W. W., Krahl, M. E., and Clowes, G. H. A., *J. Am. Chem. Soc.*, 64, 108, 1942.
27. Schwarz, F. P. and Wasik, S. P., *Anal. Chem.*, 48, 524, 1976.
28. Il'nitskii, A. P. and Ershova, K. P., *Vopc. Onkol.*, 16, 86, 1970.
29. Suess, M. J., *Water Res.*, 6, 981, 1972.
30. MacKay, D. and Wolkoff, A., *Environ. Sci. Technol.*, 7, 611, 1973.
31. Herbes, S. E., *Water Res.*, 11, 493, 1977.
32. Hertz, H. S., May, W. E., Wise, S. A., and Chesler, S. N., *Anal. Chem.*, 50, 429A, 1978.
33. Strup, P. E., Wilkinson, J. E., and Jones, P. W., Trace analysis of polycyclic aromatic hydrocarbons in aqueous systems using XAD-2 resin and capillary column gas chromatography — mass spectrometry analysis in carcinogenesis, in *Polynuclear Aromatic Hydrocarbones,* Jones, P. W. and Freudenthal, R. I., Eds., Raven Press, New York, 1978.
34. Basu, D. K. and Saxena, J., *Environ. Sci. Technol.*, 12, 791, 1978.
35. Acheson, M. A., Harrison, R. M., Perry, R., and Wellings, R. A., *Water Res.*, 10, 207, 1976.
36. Jones, P. W., Giammar, R. D., Strup, P. E., and Stanford, T. B., *Environ. Sci. Technol.*, 8, 806, 1976.
37. Leoni, V., Puccetti, G., and Grella, A., *J. Chromatogr.*, 106, 119, 1975.
38. Stepan, S. F. and Smith, J. F., *Water Res.*, 11, 339, 1977.
39. Gustafson, R. L. and Paleos, J., *Water and Water Pollution Handbook,* Vol. 1, Ciaccio, L. L., Ed., Marcel Dekker, New York, 1971.
40. Fritz, J. S., *Acc. Chem. Res.*, 10, 67, 1977.
41. Junk, G. A., Richard, J. J., Grieser, M. D., Witiak, D., Wihak, J. L., Arguello, M. D., Vick, R., Svec, H. J., Fritz, J. S., and Calder, G. V., *J. Chromatogr.*, 99, 745, 1974.
42. Junk, G. A., Richard, J. J., Fritz, J. S., and Svec, H. J., Resin sorption methods for monitoring selected contaminants in water, in *Identification and Analysis of Organic Pollutants in Water,* Keith, L. H., Ed., Ann Arbor Sci. Publ., Ann Arbor, MI, 1976.
43. Junk, G. A., Richard, J. J., Svec, H. J., and Fritz, J. S., *J. Am. Water Works Assoc.*, 68, 218, 1976.

44. Junk, G. A., Chriswell, C. D., Chang, R. C., Kissinger, L. D., Richard, J. J., Fritz, J. S., and Svec, H. J., *Fresenius' Z. Anal. Chem.*, 282, 331, 1976.
45. Tateda, A. and Fritz, J. S., *J. Chromatogr.*, 152, 329, 1978.
46. Van Rossum P. and Webb, R. G., *J. Chromatogr.*, 150, 381, 1978.
47. Musty, P. R. and Nickless, G., *J. Chromatogr.*, 89, 185, 1974.
48. Thurman, E. M., Malcolm, R. L., and Aiken, G. R., *Anal. Chem.*, 50, 775, 1978.
49. Smith, S. R., Cordes, H. F., and Johnson, J. H., NWC TM 3279, Naval Weapons Center, China Lake, CA, 1978.
50. Shinohara, R., Koga, M., Shinohara, F., and Hori, T., *Bunseki Kagaku*, 28, 856, 1977.
51. Thurman, E. M., Alken, G. R. and Malcolm, R. L., "Proc., 4th Joint Conference on Sensing of Environmental Pollutants," p. 166, (1978).
52. Malcolm, R. L., Thurman, E. M., and Aiken, G. R., *Trace Subst. Environ. Health*, 11, 307, 1977.
53. Gustafson, R. L., Albright, R. L., Heisler, J., Lirio, J. A., and Reid, O. T., Jr., *2nd Eng. Chem. Prod. Res. Develop.*, 7, 107, 1968.
54. Leenheer, J. A. and Huffman, E. W. D., Jr., *J. Res. U. S. Geol. Surv.*, 4, 737, 1976.
55. Leenheer, J. A. and Huffman, E. W. D., Jr., *Water Res. Invest.*, 79-4, January 1979.
56. Stuber, H. A. and Leenheer, E. W. D., Jr., *Am. Chem. Soc. Div. Fuel Chem.*, 23, 105, 1978.
57. May, W. E., Wasik, S. P., and Freeman, D. H., *Anal. Chem.*, 50, 997, 1978.
58. Burnham, A. K., Calder, G. V., Fritz, J. S., Junk, G. A., Svec, H. J., and Vick, R., *J. Am. Water Works Assoc.*, 65, 722, 1973.
59. Burnham, A. K., Calder, G. V., Fritz, J. S., Junk, G. A., Svec, H. J., and Willis, R., *Anal. Chem.*, 44, 139, 1972.
60. Stepan, S. F., Smith, J. F., Flego, U., and Reukers, J., *Water Res.*, 12, 447, 1978.
61. Brown, J. J. M., *J. Chem. Soc.*, (A), 1062, 1970.
62. Gesser, H. D., Chow, A, Davis, F. C., Uthe, J. F., and Reincke, J., *Anal. Lett.*, 4, 883, 1971.
63. Uthe, J. F., Reincke, J., and Yesser, H., *Environ. Lett.*, 3, 117, 1972.
64. Gough, K. M. and Yesser, H. D., *J. Chromatogr.*, 115, 383, 1975.
65. Navratil, J. D. and Sievers, R. E., *Am. Lab.*, 9, 33, 1977.
66. Navratil, J. D., Sievers, R. E., and Walton, H. F., *Anal. Chem.*, 49, 2260, 1977.
67. Saxena, J., Kozuchowski, J., and Basu, D. K., *Environ. Sci. Technol.*, 11, 682, 1977.
68. Basu, D. K. and Saxena, J., *Environ. Sci. Technol.*, 12, 795, 1978.
69. Braus, H., Middleton, F. M., and Walton, G., *Anal. Chem.*, 23, 1160, 1951.
70. Rosen, A. A. and Middleton, F. M., *Anal. Chem.*, 31, 1729, 1959.
71. Hoak, R. D., *Int. J. Air. Water Pollut.*, 6, 521, 1962.
72. Eichelberger, J. W. and Lichtenberg, J. J., *J. Am. Water Works Assoc.*, 62, 25, 1970.
73. Booth, R. L., Report from Sub-Committee on Applicability of Carbon Adsorption Technique to the Mission of the National Water Quality Network, January, 1963.
74. *Standard Methods for the Examination of Water and Wastewater*, 13th ed., American Public Health Assoc., New York, 1971, 259.
75. Grob, K., *J. Chromatogr.*, 84, 255, 1973.
76. Mieure, J. P. and Dietrich, M. W., *J. Chromatogr. Sci.*, 11, 559, 1973.
77. Pancirov, R. J., Searl, T. D., and Brown, R. A., Methods of analysis for polynuclear aromatic hydrocarbons in environmental samples, in Symp. Anal. Chem. of Petroleum Hydrocarbons in Marine/Aquatic Environment, American Chemical Society Meet., American Chemical Society, Washington, D.C., 1978, 855.
78. Grob, K., Grob, Jr., K., and Grob, G., *J. Chromatogr.*, 106, 299, 1975.
79. Murray, D. A. J., *J. Chromatogr.*, 177, 135, 1979.
80. Harrison, R. M., Perry, R., and Wellings, R. A., *Environ. Sci. Technol.*, 12, 1151, 1976.
81. Reichert, J. K., Kunte, H., Engelhardt, K., and Borneff, J., *Arch. Hyg. Bakteriol.*, 155, 18, 1971.
82. Borneff, J. and Fischer, R., *Arch. Hyg. Bakteriol.*, 146, 183, 1962.
83. Goldberg, M. C., DeLong, L., and Sinclair, M., *Anal. Chem.*, 45, 89, 1973.
84. Kahn, L. and Wayman, C. H., *Anal. Chem.*, 36, 1340, 1964.
85. Ahnoff, M. and Josefsson, B., *Anal. Chem.*, 46, 658, 1974.
86. Ahnoff, M. and Josefsson, G., *Anal. Chem.*, 48, 1268, 1976.
87. Eisenbeiss, F., Hein, H., Joester, R., and Naundorf, G., *Chromatogr. Newsl.*, 6, 8, 1978.
88. Ogan, K., Katz, E., and Slavin, W., *J. Chromatogr. Sci.*, 16, 517, 1978.
89. Chesler, S. N., Gump, G. H., Hertz, H. S., May, W. E., and Wise, S. A., NBS Special Publ., 464, November 1977.
90. May, W. E., Chesler, S. N., Cram, S. P., Gump, G. H., Hertz, H. S., Enagonio, D. P., and Dyszel, S. M., *J. Chromatogr. Sci.*, 13, 535, 1975.
91. Oyler, A. R., Bodenner, D. L., Welch, K. J., Liukkonen, R. J., Carlson, R. M., Kopperman, H. L., and Caple, R., *Anal. Chem.*, 50, 837, 1973.

92. Euston, C. B. and Baker, D. R., *Am. Lab.*, 9, 91, 1979.
93. Huber, J. F. K. and Becker, R. R., *J. Chromatogr.*, 142, 765, 1977.
94. Zlatkis, A. and Thiskie, H. A., *Chromatographia*, 6, 67, 1973.
95. Zlatkis, A., Bertsch, B., Bafus, D., and Lukich, H. M., *J. Chromatogr.*, 91, 379, 1974.
96. Wise, S. A., Chesler, S. N., Hertz, H. S., Hilpert, L. R., and May, W. E., Methods for polynuclear aromatic hydrocarbon analysis in the marine environment, in *Polynuclear Aromatic Hydrocarbons*, Jones, P. W. and Freudenthal, R. I., Eds., Raven Press, New York, 1978.
97. Chesler, S. N., Gump, B. H., Hertz, H. S., May, W. E., and Wise, S. A., *Anal. Chem.*, 50, 805, 1978.
98. Hertz, H. S., May, W. E., Chesler, S. N., and Gump, B. H., *Environ. Sci. Technol.*, 10, 900, 1976.
99. Barileul, G., *Aktive Kohle*, Verlag, F. Enke, Stuttgart, 1962.
100. Palamand, S. R., Markl, K. S., and Kardwich, X., *Am. Soc. Brew. Chem. Proc.*, 75, 1968.
101. Grob, K. and Zurcher, F., *J. Chromatogr.*, 117, 285, 1976.
102. Obach, E., *Organic Solvents*, Riddik, J. A. and Bunger, W. B., Ed., Wiley-Interscience, New York, 1970.
103. Versino, B., Knoppel, H., De Groot, M., Peil, A., Poelman, J., Schauenburg, H., Vissers, H., and Geiss, F., *J. Chromatogr.*, 122, 373, 1976.
104. Chang, R. C. and Fritz, J. S., *Talanta*, 25, 659, 1978.
105. Ryan, J. P. and Fritz, J. S., *J. Chromatogr. Sci.*, 16, 488, 1978.
106. Minsley, G., Detection and Identification of Trace Levels of Petroleum Products in Aquatic Environment, M. S. thesis, University of Wisconsin, Madison, Wisconsin, 1973.
107. Veith, G. D. and Kiwus, L. M., *Bull. Environ. Contam. Toxic*, 17, 631, 1977.
108. Lysyj, I. and Russel, E. C., Determination of aromatic and PAH content of oily wastewater, in *Symposium Analytical Chemistry of Petroleum Hydrocarbons in Aquatic Environment*, American Chemical Society, Washington, D.C., 1978, 910.
109. Grimmer, G., Bohnke, H., and Borwitzky, H., *Fresenius' Z. Anal. Chem.*, 289, 91, 1978.
110. Natusch, D. F. S. and Tomkins, B. A., *Anal. Chem.*, 50, 1429, 1978.
111. Hoffman, D. and Wynder, E. L., *Anal. Chem.*, 32, 295, 1960.
112. Haenni, E. D., Howard, J. W., and Loe, F. L., Jr., *J. Assoc. Off. Anal. Chem.*, 45, 68, 1962.
113. Sorrell, R. K., Dressman, R. C., and McFarren, E. F., High Pressure Liquid Chromatography for the Measurement of Polynuclear Aromatic Hydrocarbons in Water, paper presented at the Water Quality Control Conf., Kansas City, Mo., December 1977.
114. Sorrell, R. K., private communication, February, 1980.
115. Giger, W. and Blumer, M., *Anal. Chem.*, 46, 1663, 1974.
116. Giger, W. and Schaffner, C., *Anal. Chem.*, 50, 243, 1978.
117. Shackelford, J. and Keith, L. H., Frequency of Organic Compounds Identified in Water, EPA-600/476-062.
118. Touchstone, J. C. and Dobbins, M. F., *Practice of Thin-Layer Chromatography*, John Wiley & Sons, New York, 1978.
119. Nielsen, T., *J. Chromatogr.*, 170, 147, 1979.
120. Lankmayr, E. P. and Muller, K., *J. Chromatogr.*,170,139 1979.
121. Daisey, J. M. and Leyko, M. A., *Anal. Chem.*, 51, 24, 1979.
122. Gearing, J. N., Gearing, P. J., Lytle, T. F., and Lytle, J. S., *Anal. Chem.*, 50, 1833, 1978.
123. Borneff, J., Fate of carcinogens in aquatic environments, in *Advances in Environmental Science and Technology*, Vol. 8, Pitts, J. N., Metcalf, R. C., and Lloyd, A. C., Eds., John Wiley & Sons, New York, 1977.
124. Dunn, B. P., *Environ. Sci. Technol.*, 10, 1018, 1976.
125. Snyder, L. R., *J. Chromatogr.*, 12, 488, 1963.
126. Hurtubise, R. J., Phillip, J. D., and Skar, G. T., *Anal. Chim. Acta*,. 101, 333, 1978.
127. Pierce, R. C. and Katz, M., *Anal. Chem.*, 47, 1743, 1975.
128. Katz, M., Sakuma, T., and Ho, A., *Environ. Sci. Technol.*, 12, 909, 1978.
129. Mainwaring, S. and McGuirk, J. M., *Anal. Tech. Determination Air Pollut.*, 84, 1977.
130. Bories, J., *J. Chromatogr.*, 130, 387, 1977.
131. Shiraishi, Y., Yamashita, T., and Shirotori, T., *Eisei Kagaku*, 23, 310, 1977.
132. Scholz, I. and Altmann, H. J., *Fresenius' Z. Anal. Chem.*, 240, 81, 1968.
133. Woidich, H., Pfannhauser, W., Blaicher, G., and Tiefenbacher, K., *Chromatographia*, 10, 140, 1977.
134. Kaschari, D. T. and Reiter, R., *Fresenius' Z. Anal. Chem.*, 292, 141, 1978.
135. Pre-Symposium Notes for the 4th Int. Symp. Polycyclic Aromatic Hydrocarbons, Columbus, Ohio, October 1979.
136. Snyder, L. R., Dolan, J. W., and Gant, J. R., *J. Chromatogr.*, 165, 3, 1979.
137. Dolan, J. W., Gant, J. R., and Snyder, L. R., *J. Chromatogr.*, 165, 31, 1979.

138. Grushka, E. and Kitka, E. J., *Anal. Chem.*, 49, 1004A, 1977.
139. Wise, S. A., and Chesler, S. N., Hertz, H. S., May, W. E., Bonnett, W. J., Brown, J. M., Guenther, F. R., Hilpert, L. R., and Parris, R. M., Normal and Reverse Phase Liquid Chromatographic Separations of Polycyclic Aromatic Hydrocarbons, paper presented at the 4th Int. Symp. PAH, Columbus, OH., October 1979.
140. Ogan, K., Katz, E., and Slavin, W., *Anal. Chem.*, 51, 1315, 1979.
141. Ogan, K., Katz, E., and Slavin, W., *Chromatogr. Newsl.*, 7, 9, 1979.
142. Das, B. S. and Thomas, G. H., *Anal. Chem.*, 50, 967, 1978.
143. Fox, M. A. and Staley, S. W., *Anal. Chem.*, 48, 992, N1976.
144. Dong, M., Locke, D. C., and Ferrand, E., *Anal. Chem.*, 48, 368, 1976.
145. Krstulovic, A. M., Rosie, D. M., and Brown, P. R., *Anal. Chem.*, 48, 1383, 1976.
146. Krstulovic, A. M., Rosie, D. M., and Brown, P. R., *Am. Lab.*, 9, 11, 1977.
147. Kasiske, D., Klinkmuller, K. D., and Sonneborn, M., *J. Chromatogr.*, 149, 703, 1978.
148. Chmielowiec, J. and Sawatzky, H., *J. Chromatogr., Sci.*, 17, 245, 1979.
149. Hunt, D. C., Wild, P. J., and Crosby, N. T., *J. Chromatogr.*, 130, 320, 1977.
150. Hunt, D. C., Wild, P. J., and Crosby, N. T., *Water Res.*, 12, 643, 1978.
151. Crathorne, B. and Fielding, M., *Proc. Anal. Div. Chem. Soc.*, 15, 155, 1978.
152. Lockmuller, C. H. and Amoss, C. W., *J. Chromatogr.*, 108, 85, 1975.
153. Vespalec, R., *J. Chromatogr.*, 118, 47, 1976.
154. Popl, M., Dolansky, V., and Mostecky, J., *J. Chromatogr.*, 117, 117, 1976.
155. Popl, M., Dolansky, V., and Mostecky, J., *J. Chromatogr.*, 91, 649, 1974.
156. Popl, M., Dolansky, V., and Coupek, J., *J. Chromatogr.*, 130, 195, 1977.
157. Wise, S. A., Chesler, S. N., Hertz, H. S., Hilpert, L. R., and May, W. E., *Anal. Chem.*, 49, 2306, 1977.
158. Slavin, W., William, A. T. R., and Adams, R. F., *J. Chromatogr.*, 134, 121, 1977.
159. Christensen, R. G. and May, W. E., *J. Liq. Chromatogr.*, 1, 385, 1978.
160. Arpino, P. J. and Guiochon, G., *Anal. Chem.*, 51, 682A, 1979.
161. Van Deemter, J. J., Zuiderweg, F. J., and Klinkenberg, A., *Chem. Eng. Sci.*, 5, 271, 1956.
162. Analabs, Inc., Gas-Thin Layer-Liquid Chromatography-Chemicals and Accessories, Catalog 18, 1976.
163. Wells, D. E., *Dyn. Mass. Spectrom.*, 5, 89, 1978.
164. Thomas, R. S., Lao, R. C., Wang, D. T., Robinson, D., and Sakuma, T., *Polynuclear Aromatic Hydrocarbons,* Jones, P. W. and Freudenthal, R. I., Eds., Raven Press, New York, 1978, 9.
165. Lao, R. C., Thomas, R. S., and Monkman, J. L., *Polynuclear Aromatic Hydrocarbons,* Vol. 1, Jones, P. W. and Freudenthal, R. I., Eds., Raven Press, New York, 1976, 271.
166. Lao, R. C., Thomas, R. S., Oja, H., and Dubois, L., *Anal. Chem.*, 45, 908, 1973.
167. Lao, R. C., Thomas, R. S., and Monkman, J. L., *J. Chromatogr.*, 112, 681, 1975.
168. Strosher, M. T. and Hodgson, G. W., *ASTM Special Tech. Publ.*, 573, 259, 1975.
169. Strand, J. W. and Andren, A. W., *Anal. Chem.*, 50, 1508, 1978.
170. Janini, G. M., Johnston, K., and Zielinski, W. L., Jr., *Anal. Chem.*, 47, 670, 1975.
171. Janini, G. M., Muschik, G. M., and Zielinski, W. L., Jr., *Anal. Chem.*, 48, 809, 1976.
172. Janini, G. M., Muschik, G. M., Schroer, J. A., and Zielinski, W. L., Jr., *Anal. Chem.*, 48, 1879, 1976.
173. Janini, G. M., Shailch, B., and Zielinski, W. L., Jr., *J. Chromatogr.*, 132, 136, 1977.
174. Novotny, M., *Anal. Chem.*, 50, 16A, 1978.
175. Grob, K. and Grob, G., *Identification and Analysis of Organic Pollutants in Water,* Keith, L. H., Ed., Ann Arbor Science Publishers, Ann Arbor, Michigan, 1975, 75.
176. Cram, S. P. and Yang, F. S., Symp. Anal. Chem. of Petroleum Hydrocarbons in Marine/Aquatic Environment, American Chemical Society Meet., American Chemical Society, Washington, D.C., 1978, 845. Preprints.
177. Lee, M. L., Vassilaros, D. L., Phillips, L. V., Hercules, D. M., Asumaya, H., Jorgenson, J. W., Maskarinec, M. P., Novotny M., *Anal. Lett.*, 12(A2), 191, 1979.
178. Meuser, J. M., Bjorseth, A., Moore, F. R., Strup, P. E., and Wilkinson, J. E., Abstract from 4th Int. Symp. Polynuclear Aromatic Hydrocarbons, October 2—4, 1979, Columbus, Ohio.
179. Schomburg, G., Dielmann, R., Borwitzky, H., and Husmann, H., *J. Chromatogr.*, 167, 337, 1978.
180. Borwitzky, H. and Schomburg, G., *J. Chromatogr.*, 170, 99, 1979.
181. Schomburg, G., Husmann, H., and Borwitzky, H., *Chromatographia,* 12, 651, 1979.
182. Blomburg, L. and Wannman, T., *J. Chromatogr.*, 148, 379, 1978.
183. Dandeneau, R., Bente, P., Rooney, T., and Hiskes, R., *Am. Lab.*, 11, 61, 1979.
184. Lee, M. L., Vassilaros, D. L., White, C. M., and Novotny, M., *Anal. Chem.*, 51, 768, 1979.
185. Bjørseth, A. L., *Anal. Chim. Acta,* 94, 21, 1977.
186. Bjørseth, A., *Polynuclear Aromatic Hydrocarbons,* Jones, P. W. and Freudenthal, R. I., Eds., Raven Press, New York, 1978, 75.

187. Bjørseth, A. and Eklund, G., *Anal. Chim. Acta,* 105, 119, 1979.
188. Bjørseth, A., *Polynuclear Aromatic Hydrocarbons,* Jones, P. W. and Leber, P., Eds., Ann Arbor Science Publisher, Ann Arbor, MI, 1979, 371.
189. Lao, R. C. and Thomas, R. S., *Polynuclear Aromatic Hydrocarbons* Jones, P. W. and Leber, P., Ann Arbor Science Publishers, Ann Arbor, MI, 1979, 429.
190. Blomberg, L., Buijlen, J., Gaudzik, J., and Wannmann, T., *Chromatographia,* 11, 521, 1978.
191. Stenberg, U., Alsoberg, T., Blomberg, L., and Wannamann, T., *Polynuclear Aromatic Hydrocarbons,* Jones P. W. and Leber, P., Eds., Ann Arbor Science Publishers, Ann Arbor, MI, 1979, 313.
192. Blomberg, L. and Wannman, T., *J. Chromatogr.,* 168, 81, 1979.
193. Manufacturers chromatogram, J. & W Scientific, Inc., Orangevale, CA, 1980.
194. Maskarinec, M. P. and Olerich, G., *Anal. Chem.,* 52, 588, 1980.
195. Grob, F. L., Ed., *Modern Practice of Gas Chromatography,* John Wiley & Sons, New York, 1977.
196. Burchill, P., Herod, A. A., and James, R. G., *Polynuclear Aromatic Hydrocarbons,* Jones, P. W. and Freudenthal, R. I., Eds., Raven Press, New York, 1978, 35.
197. Radecki, A., Lamparczyk, H., Grzybowski, J., and Kalkiewicz, J., *J. Chromatogr.,* 150, 527, 1978.
198. Freed, D. J. and Faulkner, L. R., *Anal. Chem.,* 44, 1194, 1972.
199. Robinson, J. W. and Goodbread, J. P., *Anal Chim. Acta,* 66, 239, 1973.
200. Caddy, D. E. and Meek, D. M., *Proc. Anal. Div Chem. Soc.,* 13, 45, 1976.
201. Cooney, R. P. and Winefordner, J. D., *Anal Chem.,* 49, 1057, 1977.
202. Driscoll, J. N. and Spaziani, F. F., *Res. Dev.,* 27, 50, 1976.
203. Driscoll, J. N., *Am. Lab.,* 8, 71, 1976.
204. Driscoll, J. N., *J. Chromatogr.,* 134, 49, 1977.
205. Driscoll, J. N., Ford, J., Jaramillo, L., Becker, J. H., Hewitt, G., Marshall, J. K., and Onishuk, F., *Am. Lab.,* 10, 147, 1978.
206. HNU Systems Inc., Form PID/210778.
207. Driscoll, J. N., Marshall, J. K., Jaramillo, L. F., Hewitt, G., and Alongi, V., *Am. Lab.,* 12, 84, 1980.
208. Driscoll, J. N., Ford, J., Jaramillo, L. F., and Gruber, E. T., *J. Chromatogr., 158, 171, 1978.*
209. Hester, N. H. and Meyer, R. A., *Environ. Sci. Technol.,* 13, 107, 1979.
210. McFadden, W. H., *J. Chromatogr. Sci.,* 17, 2, 1979.
211. DeBrauw, M. C. T., *J. Chromatogr.,* 165, 207, 1979.
212. Fenselau, C., *Anal. Chem.,* 49, 563A, 1977.
213. Clerc, J. T., Kutter, M., Reinhard, M., and Schwarzenbach, R., *J. Chromatogr.,* 123, 271, 1976.
214. Naegeli, P. R. and Clerc, J. T., *Anal. Chem.,* 46, 739A, 1974.
215. McGuire, J. M., Carter, M. H., and Alford, A. L., Identification and Analysis of Organic Pollutants in Water, Keith, L. H., Ed., Ann Arbor Science Publisher, Ann Arbor, MI, 1975, 175.
216. Finnigan, R. E. and Knight, J. B., *Identification and Analysis of Organic Pollutants in Water,* Keith, L. H., Ed., Ann Arbor Science Publisher, Ann Arbor, MI, 1975, 185.
217. Benoit, F. M., Lebel, G. L., and Williams, D. T., *Int. J. Environ. Anal. Chem.,* 6, 277, 1979.
218. Finnigan, R. E., Hoyt, D. W., and Smith, D. E., *Environ. Sci. Technol.,* 13, 534, 1979.
219. Budde, W. L. and Eichelberger, J. W., *Anal. Chem.,* 51, 567A, 1979.
220. Vo-Dinh, T., Walden, G. L., and Winefordner, J. D., *Anal. Chem.,* 49, 1126, 1977.
221. Bower, E. L., and Winefordner, J. D., *Anal. Chim. Acta,* 102, 1, 1978.
222. Vo-Dinh, T., Bower, E. L., and Winefordner, J. D., *Talanta,* 24, 146, 1977.
223. Gammage, R. B., Vo-Dink, T., Hawthorne, A. R., Thorngate, J. H., and Parkinson, W. W., *Polynuclear Aromatic Hydrocarbons,* Jones, P. W. and Freudenthal, R. I., Eds., Raven Press, New York, 1978, 155.
224. Lloyd, J. B. F. and Evett, I. W., *Anal. Chem.,* 49, 1710, 1977.
225. Vo-Dinh, T., *Anal. Chem.,* 50, 396, 1978.
226. Vo-Dinh, T., Gammage, R. B., Hawthorne, A. R., Thorngate, J. H., *Environ. Sci. Technol.,* 12, 1297, 1978.
227. Vo-Dinh, T. and Gammage, R. B., *Anal. Chem.,* 50, 2054, 1978.
228. Latz, H. W., Ullman, A. H., and Winefordner, J. D., *Anal. Chem.,* 50, 2148, 1978.
229. Wehry, E. L., Mamantov, G., Kemmerer, R. R., Stroupe, R. C., Tokousbalides, P. T., Hinton, E. R., Hembree, D. M., Dickinson, R. B., Garrison, A. A., Bilotta, P. V., and Gore, R. R., *Polynuclear Aromatic Hydrocarbons,* Jones P. W. and Freudenthal, R. I., Eds., Raven Press, New York, 1978, 193.
230. Mamantov, G., Wehry, E. L., Kemmerer, R. R., and Hinton, E. R., *Anal. Chem.,* 49, 86, 1977.
231. Stroupe, R. C., Tokousbalides, P., Dickinson, R. B., Wehry, E. L., and Mamantov, G., *Anal. Chem.,* 49, 701, 1977.
232. Tokousbalides, P., Wehry, E. L., and Mamantov, G., *J. Phys. Chem.,* 81, 1769, 1978.
233. Tokousbalides, P., Hinton, E. R., Dick inson, R. B., Bilotta, R. V., Wehry, E. L., and Mamantov, G., *Anal. Chem.,* 50, 1189, 1978.

234. Kirkbright, G. F. and Delima, C. G., *Analyst*, 99, 338, 1974.
235. Cause B. S., Kirkbright, G. F., and DeLima, C. G., *Analyst*, 101, 367, 1976.
236. Farooq, R. and Kirkbright, G. F., *Analyst*, 101, 566, 1976.
237. Richardson, J. H. and Ando, M. E., *Anal. Chem.*, 49, 955, 1977.
238. Talmi, Y., Baker, D. C., Jademec, J. R., and Saner, W. A., *Anal. Chem.*, 50, 936A, 1978.
239. Johnson, D. W., Callis, J. B., and Christian, G. D., *Anal. Chem.*, 49, 747A, 1977.
240. O'Haver, T. C., Modulation and derivative techniques in luminescence spectrometry: approaches to increased analytical selectivity, in *Modern Fluorescence Spectroscopy*, Wehry, E. L., Ed., Vol. 1, Plenum Press, New York, 1976, 65.
241. O'Haver, T. C., *Anal. Chem.*, 51, 91A, 1979.
242. Hawthorne, A. R. and Thorngate, J. H., Report 1977, Conf-770376-1; C. A., 88, 182091C, 1978.
243. Hawthorne, A. R. and Thorngate, J. H., *Appl. Opt.*, 17, 724, 1978.
244. O'Haver, T. C. and Parks, W. M., *Anal. Chem.*, 46, 1886, 1974.

INDEX

A

B

C

D

E

F

G

H

I

K

L

M

N

O

P

Q

R

S

T